Pennant Fever

Permutations, Combinations, and the Binomial Distribution

Teacher's Guide

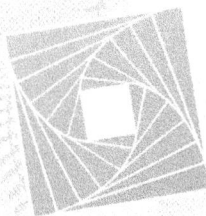

This material is based upon work supported by the National Science Foundation under award numbers ESI-9255262, ESI-0137805, and ESI-0627821. Any opinions, findings, and conclusions or recommendations expressed in this publication are those of the authors and do not necessarily reflect the views of the National Science Foundation.

Key Curriculum
1150 65th Street
Emeryville, California 94608
email: editorial@keypress.com
www.keycurriculum.com

First Edition Authors

Dan Fendel, Diane Resek, Lynne Alper, and Sherry Fraser

Contributors to the Second Edition

Sherry Fraser, Jean Klanica, Brian Lawler, Eric Robinson, Lew Romagnano, Rick Marks, Dan Brutlag, Alan Olds, Mike Bryant, Jeri P. Philbrick, Lori Green, Matt Bremer, Margaret DeArmond

Project Editor

Josephine Noah

Consulting Editor

Mali Apple

Editorial Assistant

Emily Reed

Professional Reviewer

Rick Marks, Sonoma State University

Math Checker

Carrie Gongaware

Production Director

Christine Osborne

Production Editor

Andrew Jones

Executive Editor

Josephine Noah

Mathematics Product Manager

Elizabeth DeCarli

Publisher

Steven Rasmussen

Contents

Introduction

Pennant Fever Unit Overview

Intent

In this unit, students explore probability, combinatorial coefficients, and the binomial theorem.

Mathematics

Here is a summary of the main concepts and skills that students will encounter and practice in this unit:

Probability and statistics

- Developing a mathematical model for a complex probability situation
- Using area diagrams and tree diagrams to find and explain probabilities
- Using a simulation to understand a situation, to help analyze probabilities, and to support a theoretical analysis
- Finding expected value
- Finding and using probabilities for sequences of events
- Using specific problem contexts to develop the binomial distribution, and finding a formula for the associated probabilities
- Using probability to evaluate null hypotheses

Counting principles

- Developing systematic lists for complex situations
- Using the multiplication principle for choosing one element from each of several sets
- Defining and using the concepts of permutation and combination
- Understanding and using standard notation for counting permutations and combinations
- Developing formulas for the permutation and combinatorial coefficients

Pascal's triangle and combinatorial coefficients

- Finding patterns and properties within Pascal's triangle
- Recognizing that Pascal's triangle consists of combinatorial coefficients
- Explaining the defining pattern and other properties of Pascal's triangle using the meaning of combinatorial coefficients
- Developing and explaining the binomial theorem

Other concepts and skills are developed in connection with Problems of the Week.

Progression

It's almost the end of the baseball season, and only two teams still have a chance to win the pennant: the Good Guys, with 96 wins and 59 losses; the Bad Guys, with 93 wins and 62 losses.

These two teams each have seven games left, none of which are against each other. What are the chances that the Good Guys will win the pennant?

This scenario forms the central problem of this unit, in which students will explore probability, area and tree diagrams, permutations and combinations, Pascal's triangle, the binomial distribution, the binomial theorem, and statistical reasoning.

After an initial exploration of the central unit problem, students review and solidify the use of area and tree diagrams as a means for analyzing situations involving sequences of events. They use these models to understand that the probability of a sequence of independent events is the product of their probabilities. The well-known "birthday problem" is one of the problems they solve in exploring this question:

> *What is the minimum number of people needed so that the probability of having at least one brithday match is greater than $\frac{1}{2}$?*

(The answer surprises most people.)

An important feature of the baseball pennant race is that certain overall outcomes can happen in more than one way. For example, students see that there are seven different sequences of wins and losses that can give the Good Guys a record of six wins and one loss in their final seven games. This aspect of the unit problem leads to a study of other "counting problems" and ultimately to the basic principles of permutations and combinations.

Students compile lists and use a variety of comparison arguments to develop these principles. For example, they compare the number of different three-scoop ice cream *cones* (in a 24-flavor shop) to the number of different three-scoop *bowls* of ice cream.

Students come up with formulas for counting both permutations and combinations, learn the notations $_nP_r$ and $_nC_r$, and see how to find these special numbers on a graphing calculator. (They also learn and use the $\binom{n}{r}$ notation for combinatorial coefficients.)

They see that combinatorial coefficients play a major role in probability, including situations like the baseball problem, and that the baseball problem is representative of an important family of situations. These problems are described by the binomial distribution, and students develop a general formula for the probabilities associated with this distribution.

Students also use their understanding of combinatorial coefficients and probability to tackle a variety of problems involving statistical reasoning, such as this one:

> *A high school principal is selecting a six-person subcommittee from a pool of ten adults and five students. The principal claims to be selecting people at random from the pool, but the final subcommittee consists entirely of adults. Should we suspect bias in the selection process?*

Students study Pascal's triangle in connection with their work on combinations and use the ideas in developing, understanding, and applying the binomial theorem.

Ultimately, they return to the baseball problem and apply the knowledge they have acquired to compute the desired probability. May the best team win!

The overall organization of the unit can be summarized as follows:

Play Ball!: Introduction of the central unit problem

Trees and Baseball: Development of techniques for analyzing probability for a sequence of events

The Birthday Problem: Application of probability concepts to the birthday problem

Baseball and Counting: Development of the concepts of permutations and combinations

Combinatorial Reasoning: Application of combinatorial reasoning to problems involving probability and decision-making

Pascal's Triangle: Discovery and explanation of properties of Pascal's triangle, and development of the binomial theorem

The Baseball Finale: Completion of the unit problem, portfolios, unit assessments, and reflection

Pacing Guides

50-Minute Pacing Guides (33 days)

Day	Activity	In-Class Time Estimate
1	*Play Ball!*	0
	Race for the Pennant!	35
	Introduce: *POW 10: Happy Birthday!*	15
	Homework: *Playing with Probabilities*	0
2	Discussion: *Playing with Probabilities*	15
	Discussion: *Race for the Pennant!*	30
	Homework: *Special Days*	5
3	Discussion: *Special Days*	10
	Trees and Baseball	0
	Choosing for Chores	40
	Homework: *Baseball Probabilities*	0
4	Discussion: *Choosing for Chores*	30
	Discussion: *Baseball Probabilities*	20
	Homework: *Possible Outcomes*	0
5	Discussion: *Possible Outcomes*	15
	How Likely Is All Wins?	35
	Homework: *Go for the Gold!*	0
6	Discussion: *Go for the Gold!*	30
	Discussion: *How Likely Is All Wins?*	20
	Homework: *Diagrams, Baseball, and Losing 'em All*	0
7	Discussion: *Diagrams, Baseball, and Losing 'em All*	10
	Presentations: *POW 10: Happy Birthday!*	15
	The Birthday Problem	0

	Introduce: *POW 11: Let's Make a Deal*	5
	Simulate a Deal	20
	Homework: *Day-of-the-Week Matches*	0
8	Discussion: *Day-of-the-Week Matches*	50
	Homework: *"Day-of-the-Week Matches" Continued*	0
9	Discussion: *"Day-of-the-Week Matches" Continued*	50
	Homework: *Monthly Matches*	0
10	Discussion: *Monthly Matches*	10
	The Real Birthday Problem	40
	Homework: *Six for the Defense*	0
11	*Baseball and Counting*	0
	Discussion: *Six for the Defense*	15
	And If You Don't Win 'em All?	35
	Homework: *But Don't Lose 'em All, Either*	0
12	Discussion: *And If You Don't Win 'em All?*	10
	Discussion: *But Don't Lose 'em All Either*	5
	The Good and the Bad	35
	Homework: *Top That Pizza!*	0
13	Discussion: *Top That Pizza!*	15
	Double Scoops	35
	Homework: *Triple Scoops*	0
14	Discussion: *Triple Scoops*	15
	More Cones for Johanna	35
	Homework: *Cones from Bowls, Bowls from Cones*	0
15	Discussion: *Cones from Bowls, Bowls from Cones*	15
	Bowls for Jonathan	35
	Homework: *At the Olympics*	0
16	Discussion: *At the Olympics*	30

	Presentations: *POW 11: Let's Make a Deal*	15
	Introduce: *POW 12: Fair Spoons*	5
	Homework: *Which is Which?*	0
17	Discussion: *Which is Which?*	10
	Formulas for $_nP_r$ and $_nC_r$	40
	Homework: *Who's on First?*	0
18	Discussion: *Who's on First?*	10
	Five for Seven	40
	Homework: *More Five for Sevens*	0
19	Discussion: *More Five for Sevens*	20
	Combinatorial Reasoning	0
	What's for Dinner?	30
	Homework: *All or Nothing*	0
20	Discussion: *All or Nothing*	25
	Discussion: *What's for Dinner*	25
	Homework: *The Perfect Group*	0
21	Presentations: *POW 12: Fair Spoons*	25
	Discussion: *The Perfect Group*	15
	Introduce: *POW 13: And a Fortune, Too!*	10
	Homework: *Feasible Combinations*	0
22	Discussion: *Feasible Combinations*	10
	About Bias	40
	Homework: *Binomial Powers*	0
23	Discussion: *Binomial Powers*	10
	Don't Stand for It	40
	Homework: *Stop! Don't Walk!*	0
24	Discussion: *Stop! Don't Walk!*	10
	Pascal's Triangle	0
	Pascal's Triangle	30
	Homework: *Hi, There!*	10
25	Discussion: *Pascal's Triangle*	30

	Discussion: *Hi, There!*	20
	Homework: *Pascal and the Coefficients*	0
26	Discussion: *Pascal and the Coefficients*	10
	Combinations, Pascal's Way	40
	Homework: *Binomials and Pascal–Part I*	0
27	Discussion: *Binomials and Pascal–Part I*	10
	Discussion: *Combinations, Pascal's Way*	40
	Homework: *Binomials and Pascal–Part II*	0
28	Discussion: *Binomials and Pascal–Part II*	50
	Homework: *A Pascal Portfolio*	0
29	*The Baseball Finale*	0
	"Race for the Pennant!" Revisited	45
	Homework: *Graphing the Games*	5
30	Discussion: *"Race for the Pennant!" Revisited*	25
	Discussion: *Graphing the Games*	25
	Homework: *Binomial Probabilities*	0
31	Discussion: *Binomial Probabilities*	10
	Presentations: *POW 13: And a Fortune, Too!*	35
	Homework: *"Pennant Fever" Portfolio*	5
32	*In-Class Assessment"*	30
	Homework: *Take-Home Assessment*	20
33	Exam Discussion	35
	Unit Reflection	15

90-Minute Pacing Guide (22 days)

Day	Activity	In-Class Time Estimate
1	*Play Ball!*	0
	Race for the Pennant!	70
	Introduce: *POW 10: Happy Birthday!*	20
	Homework: *Playing with Probabilities*	0
2	Discussion: *Playing with Probabilities*	15
	Trees and Baseball	0
	Choosing for Chores	70
	Homework: *Special Days*	5
3	Discussion: *Special Days*	10
	Baseball Probabilities	50
	Possible Outcomes	30
	Homework: *Go for the Gold!*	0
4	Discussion: *Go for the Gold!*	25
	Discussion: *Possible Outcomes*	15
	How Likely Is All Wins?	50
	Homework: *POW 10: Happy Birthday!*	0
5	*Diagrams, Baseball, and Losing 'em All*	40
	Presentations: *POW 10: Happy Birthday!*	15
	The Birthday Problem	0
	Introduce: *POW 11: Let's Make a Deal*	5
	Simulate a Deal	30
	Homework: *Day-of-the-Week Matches*	0
6	Discussion: *Day-of-the-Week Matches*	35
	"Day-of-the-Week Matches" Continued	55
	Homework: *Monthly Matches*	0
7	Discussion: *Monthly Matches*	10

14	Discussion: *All or Nothing*	25
	The Perfect Group	35
	About Bias	30
	Homework: *Feasible Combinations*	0
15	Discussion: *Feasible Combinations*	10
	Discussion: *About Bias*	10
	Presentations: *POW 12: Fair Spoons*	20
	Introduce: *POW 13: And a Fortune, Too!*	10
	Don't Stand for It	40
	Homework: *Binomial Powers*	0
16	Discussion: *Binomial Powers*	10
	Pascal's Triangle	0
	Pascal's Triangle	60
	Homework: *Hi, There!*	20
17	Discussion: *Hi, There!*	20
	Pascal and the Coefficients	35
	Combinations, Pascal's Way	35
	Homework: *Stop! Don't Walk!*	0
18	Discussion: *Stop! Don't Walk!*	10
	Discussion: *Combinations, Pascal's Way*	40
	Binomials and Pascal–Part I	40
	Homework: *Binomials and Pascal–Part II*	0
19	Discussion: *Binomials and Pascal–Part II*	45
	The Baseball Finale	0
	"Race for the Pennant!" Revisited	45
	Homework: *A Pascal Portfolio*	0
20	Discussion: *A Pascal Portfolio*	10
	Discussion: *"Race for the Pennant!" Revisited*	25
	Graphing the Games	55
	Homework: *Binomial Probabilities*	0
21	Discussion: *Binomial Probabilities*	10

	Presentations: *POW 13: And a Fortune, Too!*	30
	In-Class Assessment	30
	Homework: *Take-Home Assessment*	20
22	Exam Discussion	45
	Unit Reflection	15
	Homework: *"Pennant Fever" Portfolio*	30

Materials and Supplies

All IMP classrooms should have a set of standard supplies, described in the section "Materials and Supplies for the IMP Classroom" in *A Guide to IMP*. You'll also find a comprehensive list of materials needed for all Year 3 units in the section "Materials and Supplies for Year 3" in the *Year 3 Teacher's Guide* general resources.

Listed here are the supplies needed for this unit. Also available are general and activity-specific blackline masters, for transparencies or for student worksheets, in the "Blackline Masters" section in *Pennant Fever* Unit Resources.

Pennant Fever Materials

Optional: Copies of a calendar for the current month (1 per group)

A variety of materials that groups are likely to request for conducting a simulation, such as colored cubes, coins, paper bags

Assessing Progress

Pennant Fever concludes with two formal unit assessments. In addition, there are many opportunities for more informal, ongoing assessments throughout the unit. For more information about assessment and grading, including general information about the end-of-unit assessments and how to use them, consult the *Year 3: A Guide to IMP* resource.

End-of-Unit Assessments

This unit concludes with in-class and take-home assessments. The in-class assessment is intentionally short so that time pressures will not affect student performance. Students may use graphing calculators and their notes from previous work when they take the assessments. You can download unit assessments from the *Pennant Fever* Unit Resources.

Ongoing Assessment

One of the primary tasks of the classroom teacher is to assess student learning. Although the assigning of course grades may be part of this process, assessment more broadly includes the daily work of determining how well students understand key ideas and what level of achievement they have attained on key skills, in order to provide the best possible ongoing instructional program for them.

Students' written and oral work provides many opportunities for teachers to gather this information. Here are some written assignments and oral presentations to monitor especially carefully for insight into student progress:

- *Baseball Probabilities*
- *How Likely Is All Wins?*
- *Monthly Matches*
- *Cones from Bowls, Bowls from Cones*
- *Who's on First?*
- *About Bias*
- *Race for the Pennant Revisited*

Discussion of Unit Assessments

Ask for volunteers to explain their work on each of the problems. Encourage questions and alternate explanations from other students.

In-Class Assessment

On Question 1, the key is seeing that Janice is not yet deciding the order in which she will read the five books, so the number of choices is $_{16}C_5$ (and not $_{16}P_5$). In contrast, on Question 2, Lori is deciding on the order of her five biking trips, so the number of choices is $_{10}P_5$.

Question 3 is more complex because it involves two choices for each day. One approach is to have Lori separately choose a sequence of five pairs of shorts ($_8P_5$ choices) and a sequence of five shirts ($_{12}P_5$ choices). Altogether, that gives $_8P_5 \cdot {}_{12}P_5$ choices (using the multiplication principle discussed after *Possible Outcomes* and used in *Who's on First?*).

Another approach to Question 3 is to see that Lori has $8 \cdot 12$ choices for Monday, then $7 \cdot 11$ choices for Tuesday, and so on. This leads to the same answer.

Take-Home Assessment

There are several ways to explain Question 1. One approach is to see that there are $4 \cdot \binom{13}{5}$ possible five-card hands in which the cards are all of the same suit. (Such a hand is called a *flush* in poker.) Because there are $\binom{52}{5}$ five-card hands altogether, the probability of getting five cards of the same suit is

$$\frac{4 \cdot \binom{13}{5}}{\binom{52}{5}}$$

A very different approach is to view the cards as dealt one at a time. Once the first card is dealt, you can then find the probability for each subsequent card that it is the same suit as the first. In this approach, the probability is the product

$$\frac{12}{51} \cdot \frac{11}{50} \cdot \frac{10}{49} \cdot \frac{9}{48}$$

Students may find it illuminating to write the first answer showing each combinatorial coefficient as a fraction of the form $\frac{_nP_r}{r!}$ (writing $_nP_r$ itself as a product). They can then see how the cancellation of terms shows that the two answers are the same.

One shortcut to Question 2 is to first find the probability that all or all but one of the games will be rained out and then simply subtract the sum of these probabilities from 1. (This means looking at only two cases rather than the four cases of playing two games, playing three games, playing four games, or playing five games.)

For Question 3, you might have one or two students read the articles they wrote and then let other students comment or add other ideas.

Supplemental Problems

Pennant Fever contains a variety of activities at the end of the student pages that you can use to supplement the regular unit material. These activities fall roughly into two categories.

Reinforcements increase students' understanding of and comfort with concepts, techniques, and methods that are discussed in class and that are central to the unit.

Extensions allow students to explore ideas beyond those presented in the unit, including generalizations or abstractions of ideas.

The supplemental activities are presented in the *Teacher's Guide* and in the student book in the approximate sequence in which you might use them. Listed here are specific recommendations about how each activity might work within the unit. You may wish to use some of these activities, especially the later ones, after the unit is completed.

***Putting Things Together* (reinforcement)** You can assign this problem any time after the discussion of *Possible Outcomes*.

***Ring the Bells!* (reinforcement)** The situation in this activity is a bit more complex than *Go for the Gold!*, but the basic ideas are the same. This activity makes a good follow-up to that assignment.

***Programming a Deal* (extension)** Some students may enjoy writing a calculator or computer program to carry out the simulation described in *Simulate a Deal*. They can begin work on this activity after the discussion of that activity.

***Simulation Evaluation* (extension)** This activity will require students to review the use of the chi-square statistic (from the Year 2 unit *Is There Really a Difference?*). We use the null hypothesis that the two strategies are simply "equally successful" because there is no obvious choice as to the probability of success for either strategy. Students can work on this after the class compiles its results from *Simulate a Deal*.

***The Chances of Doubles* (reinforcement)** This activity provides another opportunity for students to use the technique of finding the probability of some event happening by finding the probability of that event not happening.

It is a good follow-up to the "birthday" series of problems (which end with *The Real Birthday Problem*).

Determining Dunkalot's Druthers (reinforcement) This activity is an interesting application of the ideas from the central unit problem. Students should be able to work on it successfully following the discussion of *More Five for Sevens*.

Sleeping In (reinforcement) This activity is similar to *The Perfect Group* and can be used to reinforce ideas from that assignment.

Twelve Bags of Gold Revisited (extension) Many students are not able to solve the *Twelve Bags of Gold* POW when they are Year 1 students, because it is a quite difficult problem. We include it again here so students can give it another try (or improve the quality of their write-up from the first time if they solved the problem then). You can assign this activity along with *POW 13: And a Fortune, Too!*

My Dog's Smarter Than Yours (reinforcement) This activity is similar to *Stop! Don't Walk!* Like that assignment and *Don't Stand for It,* this problem involves cumulative probabilities.

Defining Pascal (extension) This activity asks students to explore the connection between the method of extending Pascal's triangle row by row and the fact that the entries are combinatorial coefficients. Some students may work through the ideas of this assignment as part of *Combinations, Pascal's Way,* but you may want to assign this supplemental problem as a separate investigation.

Maximum in the Middle (extension) This activity also asks students to justify a relationship they explored in *Combinations, Pascal's Way.* It can be assigned after that activity, depending on how thoroughly students developed the ideas in class.

The Why's of Binomial Expansion (extension) This activity leads students to develop an explanation for why the combinatorial coefficients appear as coefficients in the expansion of binomials. It can be assigned after the formal statement of the binomial theorem following *Binomials and Pascal—Part II.*

The Binomial Theorem and Row Sums (extension) In this activity, students are asked to use the binomial theorem to prove that the row sums in Pascal's triangle are powers of 2. This can be assigned after the formal statement of the binomial theorem following *Binomials and Pascal—Part II.*

Play Ball!

Intent

The activities in *Play Ball!* introduce the unit problem.

Mathematics

The unit problem involves analysis of the probabilities for a complex situation involving many possible sequences of events. In *Play Ball!* students are introduced to this problem, they review some basic principles about probability, and they establish simplifying assumptions necessary to a complete solution.

The POW introduced in this section will lead to a later series of activities, beginning with *Day-of-the-Week Matches*, that help students learn to use tree diagrams and area models to analyze situations with multiple events.

Progression

Race for the Pennant! introduces the unit problem. *Playing with Probabilities* requires students to apply basic probability concepts. *Special Days* helps students get started on *POW 10: Happy Birthday!*

Race for the Pennant!

POW 10: Happy Birthday!

Playing with Probabilities

Special Days

Race for the Pennant!

Intent
This activity introduces the central unit problem.

Mathematics
The central unit problem is to find the probability that a team called the "Good Guys" will win the baseball pennant, given a situation in which all but two teams are out of the running.

In this activity, students identify possible outcomes of the pennant and estimate some probabilities. In the discussion following this activity, students formulate a number of simplifying assumptions necessary to a complete solution, including the assumption that the probability of each of the two remaining teams winning a game is constant and can be calculated from their records to date. For background information concerning the implications of the assumptions, read *The Binomial Distribution: An Overview for Teachers.*

Progression
Students work on the activity in groups, and then discuss as a class their work and simplifying assumptions for the unit problem.

Approximate Time
35 to 40 minutes for activity

30 minutes for discussion

Classroom Organization
Small groups, followed by whole-class discussion

Doing the Activity
Tell students that the central unit problem involves finding the probability that a certain baseball team, called the "Good Guys," will win the pennant. In the context of this unit, *winning the pennant* means "winning more games than any other team."

Neither you nor your students need to know any details about the game of baseball to work on this problem, but it will give some students a sense of ownership of the problem if they have an opportunity to share their knowledge of baseball. Try to allow some students to share their knowledge without overwhelming other students with unnecessary information.

Have students read the introduction to the problem (stop before the questions), and then be sure everyone understands the basic scenario. Here are the important details:

- The Good Guys are leading, near the end of the season, and their closest competitor is the Bad Guys.
- All other teams in the league are so far behind that they don't have a chance.
- The Good Guys will not play against the Bad Guys in any of their remaining games of the season.
- The winner of the pennant is the team that wins the most games. (The situation in major league baseball today is more complicated, because the pennant involves playoff games between division winners.)

Clarify, if necessary, that an individual game cannot end in a tie. Then have groups work on the questions in the activity.

Discussing and Debriefing the Activity

Unless you observed groups having difficulty with Question 1, begin the discussion by having several students present their groups' ideas about Questions 2 and 3. Encourage students to challenge each other's reasoning and to engage in productive debate. Be sure that presenters state their assumptions clearly.

On Question 2, students are likely to reason that the teams are each most likely to win about four of their remaining seven games, because that is roughly the proportion of games they won previously. (If students present the actual percentages of games won so far for each team, save those values for discussion later today.)

Question 3 is more complex and is the central problem of the unit. If students seem tentative, ask, **Can you say anything about the probability? Is there a range within which it must fall?** For instance, they can be fairly sure that it is greater than $\frac{1}{2}$ (and, of course, less than 1). Here are some reasons they might give for this conclusion:

- The Good Guys are now ahead of the Bad Guys, so they need fewer wins to end up ahead. Thus, they are more likely to win the pennant than the Bad Guys are.
- The Good Guys are probably a better team, because otherwise they wouldn't be ahead now. Therefore, they have a better chance of doing well in their remaining games than the Bad Guys do.

Tell students that this problem is quite complicated and that solving it will be the main focus of this unit. Acknowledge that in real life, many issues beyond mathematics would go into determining the probability that the Good Guys would win the pennant, such as which teams the Good Guys and Bad Guys were playing in their remaining games and how healthy their players were.

Other Observations

Here are some other observations that students might make as part of their work on the problem:

- If the Good Guys win five or more of their remaining games, they'll definitely win the pennant, and if they win four games, they will at least tie for the pennant.
- If the Bad Guys lose five or more of their games, they cannot win the pennant, and if they lose four games, the best they can do is get a tie.

Either of these observations can be used to refine the estimate of the probability that the Good Guys will win "more than $\frac{1}{2}$," but they are insufficient to give the exact answer.

Some Simplifying Assumptions

Devote time to establishing some simplifying assumptions for the problem—the remainder of the unit depends on them. If possible, elicit these assumptions from the class. If necessary, however, suggest them yourself.

One assumption we will make is this:

Each of the two teams has a fixed probability of winning each of their remaining games.

This means, for example, that the probability that the Good Guys will win the first of their remaining games is the same as the probability that they will win their last game.

We will also establish the fixed probabilities for each team based on their records so far. That is, we will make these two assumptions:

- The probability that the Good Guys will win any particular game is $\frac{96}{155}$.

  Actually, because this fraction is approximately .6194, we will use .62 as the Good Guys' probability of winning each of their remaining games.

- The probability that the Bad Guys will win any particular game is $\frac{93}{155}$. This is exactly .6, and we will use this value as the Bad Guys' probability of winning each of their remaining games.

Acknowledge to students that these simplifying assumptions are not necessarily correct and that perhaps they could come up with a better set of assumptions. For instance, it might be a bit more accurate to adjust each team's probability of winning a particular game each time the team wins or loses. Thus, if the Good Guys win their first game, we could say that their probability of winning their next game is $\frac{97}{156}$. However, we will keep things simpler by keeping the probabilities constant. The use of the probabilities .62 and .6 also ignores the possible variations in the strength of the opponents in the remaining games.

Try to get students to agree that the assumptions we will use are fairly reasonable. Tell them that using these assumptions will make it possible to analyze the problem completely.

Post the probabilities .62 and .6, appropriately labeled, so that students can refer to them throughout the unit.

For background information concerning the implications of these assumptions, read *The Binomial Distribution: An Overview for Teachers*.

Key Questions

Can you say anything about the probability? Is there a range within which it must fall?

The Binomial Distribution—An Overview for Teachers

The assumptions made in the discussion of *Race for the Pennant!* mean that for the two teams, the number of games each will win is determined by a **binomial distribution**. This term will be introduced to students later in the unit. The essential features of a binomial distribution are these:

- There is a sequence of events, each with two possible outcomes. These outcomes are referred to generically as *success* and *failure*.
- The probability of success is the same for each of the events (and so the probability of failure is also the same for each of the events).

In a binomial distribution, the number of events is fixed, say, n, and the probability of success for each event is usually represented by p. The *binomial variable* is the number of successes, which we will here call k. The basic task is to find, for each value of k from 0 through n, the probability that exactly k of the n events will result in success.

In the case of the unit problem, $n = 7$. For the Good Guys, we are using $p = .62$, and for the Bad Guys, we are using $p = .6$.

Ultimately (see *Binomial Probabilities*), students will develop the principle that the probability of k successes is given by the expression

$$\binom{n}{k} p^k \left(1 - p\right)^{n-k}$$

where $\binom{n}{k}$ is the **combinatorial coefficient** (or **binomial coefficient**)

representing the number of ways of choosing k objects out of a set of n objects.

POW 10: Happy Birthday!

Intent

In this activity, students use divisibility and remainders to develop a complex algorithm.

Mathematics

POW 10: Happy Birthday! asks students to develop an algorithm that can be used to determine the day of the week for the date on which anyone was born. This POW will launch a series of activities, beginning with *Special Days* and culminating in *The Real Birthday Problem*, that will strengthen students' ability to use tree diagrams and area models and to determine probabilities for multistage events.

Progression

Students will need about a week to work on this POW. Presentations should be completed before beginning *Day-of-the-Week Matches*, because students will need to know (but not reveal to the class) the day of the week on which they were born for that activity.

Approximate Time

15 to 20 minutes for introduction

1 to 3 hours for activity (at home)

15 minutes for presentations

Classroom Organization

Individuals, followed by several student presentations

Materials

Optional: Copies of a calendar for the current month

Doing the Activity

Introduce this POW by having a student read the introduction. Then ask students to share ideas from their cultures about the role (if any) that birth day-of-the-week plays in an individual's personality, or to share their reactions to the specific individuals listed for each day of the week.

Special Days will help students get started on this activity by having them look at specific examples. Schedule presentations on *POW 10: Happy Birthday!* about a week after introduction of the POW. Students will need to know the day of the week on which they were born for the discussion of *Day-of-the-Week Matches Continued*, but they should not disclose their birth days-of-the-week before then. Encourage students to find out their day of birth from their families so they can check the algorithms developed for the POW. Reassure those who can't get this information independently that they will be able to discover it through the activity.

There is no simple set of directions that will accomplish the task in this POW, but the task is fairly concrete and should be accessible on some level to most students. In part, this POW is about perseverance and attention to detail.

On the day before the POW is due, choose three students to make presentations on the following day, and give them overhead transparencies and pens to take home to use for preparing those presentations.

Discussing and Debriefing the Activity

Have three selected students make presentations. After each presentation, you may want to let the rest of the class try to apply the method described to see if it works.

Be sure that by the end of the discussion, each student has found out the day of the week of his or her own birth. But ask students to keep this information secret until the discussion following *Day-of-the-Week Matches Continued.*

Playing with Probabilities

Intent

In this activity, students review some basics about probability.

Mathematics

This activity is designed to refresh students' memories about basic principles of probability, in preparation for working on the central unit problem. Students will determine probabilities, draw area diagrams, and make up situations with specified probabilities. The subsequent discussion emphasizes that the sum of the probabilities for all the outcomes of an event must be 1.

Progression

Students work on this activity individually and then compare their results in groups and in a class discussion.

Approximate Time

20 to 25 minutes for activity (at home or in class)

15 minutes for discussion

Classroom Organization

Individuals, followed by whole-class discussion

Materials

Transparency of Playing with Probabilities blackline master

Doing the Activity

Students should be able to complete this activity independently.

Discussing and Debriefing the Activity

Have students compare answers within groups. Then have each group share selected answers with the whole class.

For Question 1, have the presenter give details on how she or he found the probability associated with each area, using a transparency of the Playing with Probabilities blackline master. Emphasize the notion of "equally likely events" by breaking up the entire area into boxes of the same size as the green (or black, on the blackline master) box in the center. (There should be 25.)

Use Question 2 to review that the probabilities of all possible outcomes of an event must sum to 1.

Have several students do brief presentations on Question 3, because there are many situations that will fit the given probabilities. Be sure presenters explain how they know that each outcome has the probability specified.

In Question 4, students must use the fact that 15% of the whole is 9 gumballs to find out what the whole was. There are many ways to reason this out, so ask several students to share their methods. One method is to divide by 3 and reason that 5% is 3 balls and 10% is 6 balls, so the total must be 60 balls. Thus, there are 15 reds (because they constitute 25 percent of the whole). Students can get the number of blues either by calculating 60% of 60 or by subtracting the number of reds and purples from the total.

Special Days

Intent

This activity will get students started working on *POW 10: Happy Birthday!*

Mathematics

Special Days asks students to figure out, using only a calendar for the current month, the day of the week for several specific dates. Remainders are used in the problem analysis.

Progression

Students work on this activity individually and then compare their results in groups and in a class discussion.

Approximate Time

5 minutes for introduction

30 minutes for activity (at home or in class)

10 minutes for discussion

Classroom Organization

Individuals, followed by whole-class discussion

Materials

Optional: Copies of a calendar for the current month

Doing the Activity

You may need to clarify for the class what is meant by "the calendar year previous to the current year." Also, you may want to distribute copies of a calendar for the current month to help students focus on what information they are allowed to use.

Discussing and Debriefing the Activity

As the purpose of this activity is simply to get students started on the POW, it will be more productive to discuss several approaches to one problem than to discuss all three problems.

Have two or three students present the first problem and focus the discussion on details of these explanations, including the leap year issue if it applies. Students probably approached the problem in a variety of ways. Be sure they realize there is no single "right way" to do these problems, even though the problems do have "right answers."

One Approach

One approach is to start by figuring out how many days have passed since January 1 of the previous year. Suppose, for instance, that the discussion is taking place on April 10. Students might note that it is 365 (or 366) days from January 1 of the previous year to January 1 of the current year. So we could add 31 days for January, 28 (or 29) for February, and 31 for March to get the total from January 1 of the previous year to April 1 of the current year, and then add 9 more days to get to April 10.

This computation shows that from January 1 of the previous year to April 10 of the current year is either 464 or 465 days. To continue this example, let's assume that either this year or last is a leap year, so the total is 465.

The next step is to find the remainder when this total number of days is divided by 7. Getting students to articulate this principle is a key purpose of this assignment.

Writing $465 = 7 \cdot 66 + 3$, we see that April 10 of this year is "3 days later in the week" than January 1 of last year. Students can apply the idea that if April 10 of this year is "3 days later in the week" than January 1 of last year, then January 1 of last year was "3 days earlier in the week" than April 10 of this year.

Finally, students can consult a current calendar to determine the day of the week for April 10 (the day of the presentation) and work backwards three days in the week to determine the day of the week for January 1 of the previous year.

Trees and Baseball

Intent

This sequence of activities develops the use of tree diagrams in calculating probabilities.

Mathematics

Tree diagrams were used in Year 1 of the IMP curriculum to develop lists of possible outcomes. *Trees and Baseball* expands this use to calculating probabilities of combinations or sequences of events, where each branch of the tree does not necessarily have the same probability. The activities build understanding by making frequent comparison between tree and area diagrams, and they begin to prepare students for calculating probabilities of situations in which construction of neither diagram will be practical.

Progression

Choosing for Chores introduces using the tree diagram to calculate the probability of a sequence of events, and *Baseball Probabilities* takes a closer look at explaining why multiplying the probabilities along a branch of the tree yields the probability for the sequence of events represented by that branch. In *How Likely Is All Wins?*, students see that this principle of multiplying probabilities holds true for finding the probability of events that are not sequential as well. The latter two activities also start students off on finding probabilities for some of the possible combinations of outcomes for the two baseball teams in the central unit problem.

Possible Outcomes introduces a different multiplication principle, this time for calculating the number of ways to pick one object from each of several sets.

Students get additional opportunities to work with and reflect upon tree diagrams, area models, and expected value in *Go for the Gold!* and *Diagrams, Baseball, and Losing 'em All*.

Choosing for Chores

Baseball Probabilities

Possible Outcomes

How Likely Is All Wins?

Go for the Gold!

Diagrams, Baseball, and Losing 'em All

Choosing for Chores

Intent

In this activity, students examine a problem involving a conditional probability and expected value.

Mathematics

Choosing for Chores gives students two problems, each involving conditional probability. The discussion looks at each of the main approaches to solving this kind of problem; students will see that tree diagrams can be used not only for developing lists of possible outcomes, but also to compute probabilities.

The activity serves as a basis for reviewing area diagrams, lists, and tree diagrams, as well as expected value.

Progression

We recommend that you discuss *Choosing for Chores* before discussing *Baseball Probabilities*, because the use of tree diagrams to find probabilities is easier to understand in *Choosing for Chores*.

This activity will be the foundation to *POW 12: Fair Spoons.*

Approximate Time

35 to 40 minutes for activity

30 to 35 minutes for discussion

Classroom Organization

Small groups or individuals, followed by whole-class discussion

Materials

Transparency of *Choosing for Chores* blackline master

Doing the Activity

No introduction is needed for *Choosing for Chores.* All groups will likely finish Part I, but it is not necessary that they all finish Part II.

Suggestions for Part I

As groups work on Part I, encourage them to use different approaches for finding the probabilities. Try to get at least one group ready to make a presentation using each of the main approaches—area diagrams, lists, and tree diagrams.

Although students used lists and tree diagrams to find probabilities in Year 1, they used them primarily in contexts involving equally likely outcomes. Students may not see right away how to create a list or a tree diagram with equally likely outcomes for this situation. You might suggest that they consider using labels with subscripts for each spoon. For instance, they might call the spoons P_1, P_2, G_1, G_2, and G_3. Refer to the discussion below for details.

Hints for Part II

Groups that get to Part II may need help recalling how to find expected value. As needed, ask, **What does expected value mean?** You might remind students of the method of finding expected value by computing the average result over a large number of trials. Help them see that for Choice 2, they will sometimes get $2 and sometimes get $6 and they need to figure out how often each will happen. It may help if they think of the bills as being pulled out of the bag one at a time.

Discussing and Debriefing the Activity

A key goal for today is to extend students' use of tree diagrams beyond developing lists of possible outcomes. In this discussion, students should see that they can also use tree diagrams to compute probabilities. Some review of the basic ideas of tree diagrams may be needed.

Part I: Wash or Dry?

Have one or two students present an analysis of the problem using each of the three general methods—area diagrams, lists, and tree diagrams. Following are some ideas about area diagrams, lists, and initial use of tree diagrams, which you may find helpful in enabling students to get the most benefit from this discussion.

Using an Area Diagram

A pair of diagrams can be used to describe the situation. For example, we can begin with this diagram for the first spoon drawn.

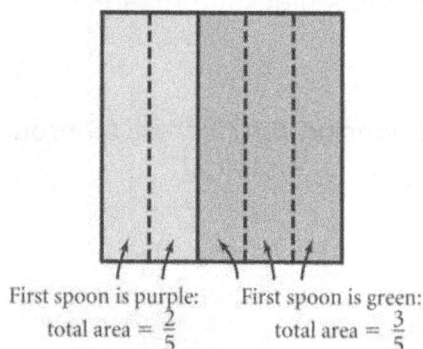

First spoon is purple:
total area = $\frac{2}{5}$

First spoon is green:
total area = $\frac{3}{5}$

The dashed lines help to show the individual spoons, and the areas of the rectangles outlined with heavy lines represent the probability of each color.

Once a spoon is drawn, however, only four spoons are left—either 1 purple and 3 green or 2 purple and 2 green. Therefore, the second-stage diagram might look like this:

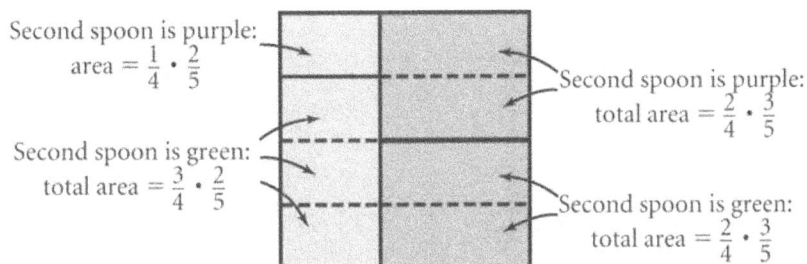

Second spoon is purple:
area $= \frac{1}{4} \cdot \frac{2}{5}$

Second spoon is purple:
total area $= \frac{2}{4} \cdot \frac{3}{5}$

Second spoon is green:
total area $= \frac{3}{4} \cdot \frac{2}{5}$

Second spoon is green:
total area $= \frac{2}{4} \cdot \frac{3}{5}$

(We used $\frac{2}{4}$ rather than $\frac{1}{2}$ in the expressions on the right side of the diagram to emphasize that there are two spoons of a given color out of a total of four spoons. Of course, students can simplify these fractions.)

Be sure students see that the area here contains four basic portions, each representing a sequence of colors drawn. As discussed for the activity problems, there are various ways to think about finding these areas, including multiplying length by width and subdividing the overall square into equal parts.

The upper-left section and the lower-right section represent a color match, and these have areas of $\frac{1}{4} \cdot \frac{2}{5}$ and $\frac{2}{4} \cdot \frac{3}{5}$, respectively. Thus, the probability of the colors matching is $\left(\frac{1}{4} \cdot \frac{2}{5}\right) + \left(\frac{2}{4} \cdot \frac{3}{5}\right)$, which is $\frac{8}{20}$, so Scott has a 40 percent chance of being the dishwasher.

If you keep the subdivision lines from the first stage of the area diagram when you go on to the second spoon, then the diagram looks like this, with the main sections labeled to show the colors of the spoons:

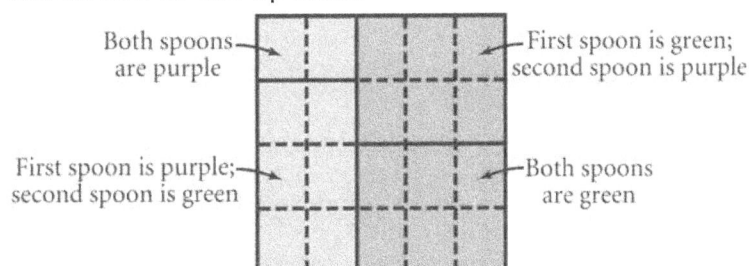

Both spoons
are purple

First spoon is green;
second spoon is purple

First spoon is purple;
second spoon is green

Both spoons
are green

As this diagram shows, 8 of the 20 equal-size sections have spoons with matching colors.

Using Lists

As suggested in the introduction to the activity, it may be helpful to label the spoons as P_1, P_2, G_1, G_2, and G_3. This labeling can lead to a list of cases:

P_1P_2	P_2P_1	G_1P_1	G_2P_1	G_3P_1
P_1G_1	P_2G_1	G_1P_2	G_2P_2	G_3P_2
P_1G_2	P_2G_2	G_1G_2	G_2G_1	G_3G_1
P_1G_3	P_2G_3	G_1G_3	G_2G_3	G_3G_2

From this list, students can simply count cases for each outcome. In particular, 8 of the 20 cases show a color match, so the probability of such a match is $\frac{8}{20}$. You can connect this approach to the area diagram by asking students to locate specific items on the list within the area diagram. For instance, they should see that the outcome G_2P_1 belongs in the upper-right portion of the previous area diagram.

Using a Tree Diagram

There are two possible approaches that may come up for illustrating this situation with a tree diagram:

- A diagram in which each spoon gets a distinct branch
- A diagram in which spoons of the same color are grouped

Students will likely take the first approach, which is preferable for the initial discussion. Although it involves a more cumbersome diagram, it provides a better tool for understanding the probabilities.

Labeling the spoons as P_1, P_2, G_1, G_2, and G_3 might lead to this tree diagram:

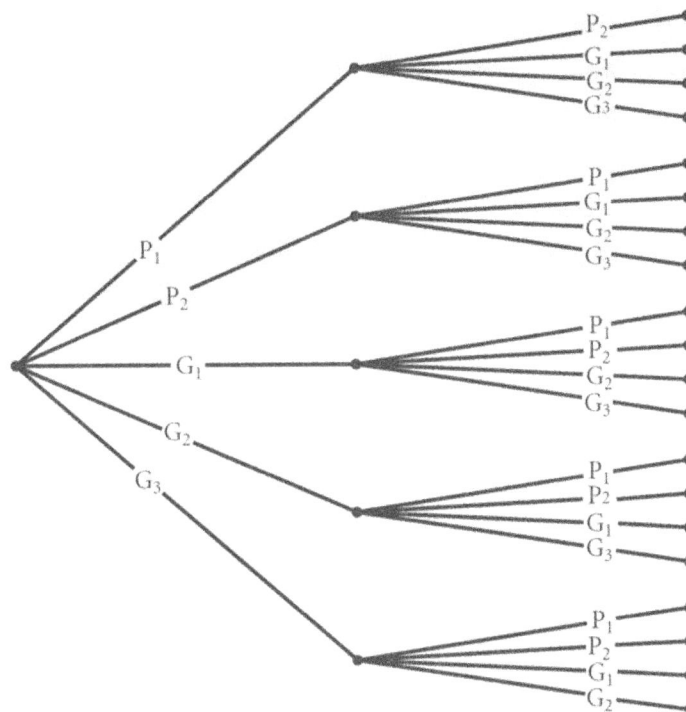

Students will probably recognize that the paths here are equally likely, and will count paths as they counted entries in the list to see that the probability of a color match is $\frac{8}{20}$.

Review the terms *branch* and *subbranch,* and introduce the term *node* for a place from which branches or subbranches emerge. Review the principle that at each node, the subbranches from that node show what can happen at that stage of the problem situation. Trace one or two paths from beginning to end to be sure students understand what such a path might represent. Also, have students relate the paths of the diagram to sections of the area diagram and to the list.

Extending the Use of Trees

Once students have created a basic tree diagram with equally likely paths, ask, **How might you label this tree diagram to show probabilities?** One standard method, illustrated below, is to label each subbranch with the probability of going on to that subbranch once you get to the preceding node.

In this situation, the five equally likely initial branches all are labeled with the probability $\frac{1}{5}$. At the end of each of these initial branches are four equally likely subbranches, so each of these is labeled with the probability $\frac{1}{4}$. For instance, the next diagram shows part of the tree and how it might be labeled.

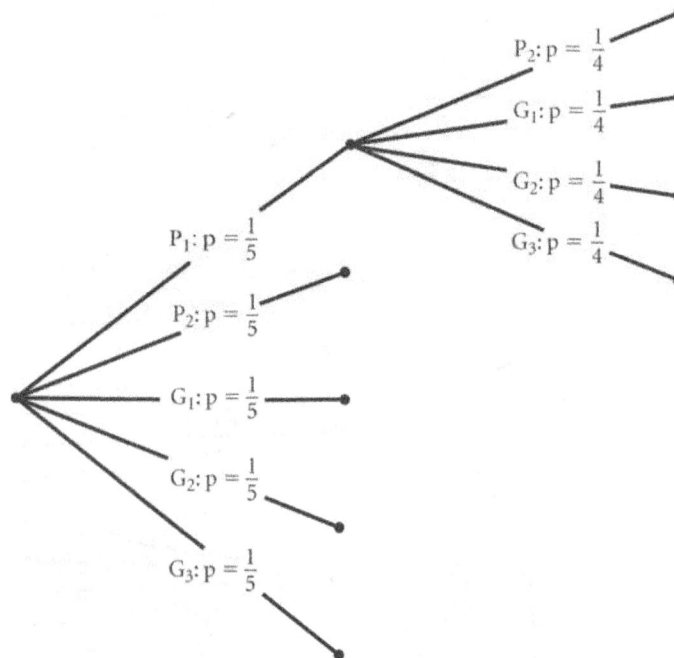

Ask, **What is the probability that the outcome will be the path P_1G_1?** Because there are five initial branches, each splitting into four subbranches, there are a total of 20 paths, each with probability $\frac{1}{20}$.

To make the connection between this probability and the probabilities for the two subbranches, ask, **How is this probability related to the probabilities for P_1 and G_1?** Bring out that $\frac{1}{20}$ is simply the product of the two individual probabilities, $\frac{1}{5}$ and $\frac{1}{4}$.

Finally, ask, **Why do you multiply these two probabilities?** Students will probably explain this using the fact that the total number of paths is $5 \cdot 4$.

Simplifying by Combining Branches

One key goal of this discussion is to help students apply the same reasoning with a simpler diagram.

Ask, **How might you combine branches to simplify the diagram?** If needed, suggest to students that they use branches to represent colors rather than spoons, and lead them to a diagram like this:

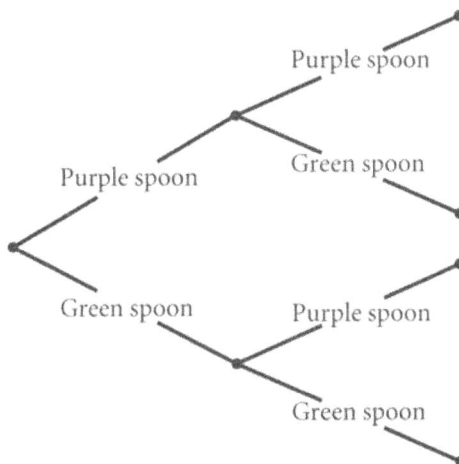

Note: The *Choosing for Chores* blackline master is a blank tree diagram of this type.

You might challenge students' understanding by suggesting that the diagram shows four paths, so, **Doesn't this show that the probability of getting a purple spoon and then a green one is** $\frac{1}{4}$**?** This should elicit the response that the outcomes shown in this tree diagram are not equally likely.

Ask, What is the probability for each branch and subbranch? This should lead to a diagram like this:

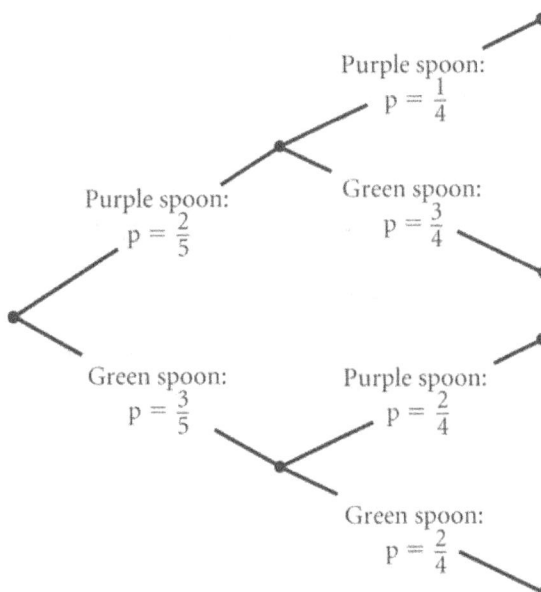

Bring out that at each stage, the probability is based on the basic definition for equally likely cases:

$$\frac{\text{number of favorable cases}}{\text{total number of cases}}$$

For instance, the initial branch labeled "green spoon" has a probability of $\frac{3}{5}$ because three of the five initial branches (which were equally likely) represented green spoons.

Note: It's helpful to initially use unsimplified fractions such as $\frac{2}{4}$, as shown in this diagram, because they more clearly reflect the basic definition of probability as a quotient of numbers of cases.

Again, ask about the probability for a particular path. For example, ask, **What is the probability of the path "purple, green"?** Students should see (perhaps going back to the original tree diagram) that the probability of the path "purple spoon, green spoon" is $\frac{6}{20}$.

The crucial element of the entire discussion is seeing that even with the simplified diagram, the probability for the path is the product of the probabilities along the branches. Again, ask students to explain this, and elicit a variety of explanations. Here are three possibilities:

- The path probabilities, $\frac{2}{5}$ and $\frac{3}{4}$, are the length and width of a rectangle in the area diagram, so their product is the area of this rectangle.

- The denominator of this product is $5 \cdot 4$, which is the total number of paths in the original tree diagram. The numerator of this product is $2 \cdot 3$, which shows the number of paths that have been combined at each stage.

- In $\frac{2}{5}$ of the cases, the first spoon will be purple, and in $\frac{3}{4}$ of those cases, the second spoon will be green. You find $\frac{3}{4}$ of $\frac{2}{5}$ by multiplying the two fractions.

You might end this discussion by suggesting that students include the probability for each path at the final node of the path. For instance, the tree diagram just discussed might look like this:

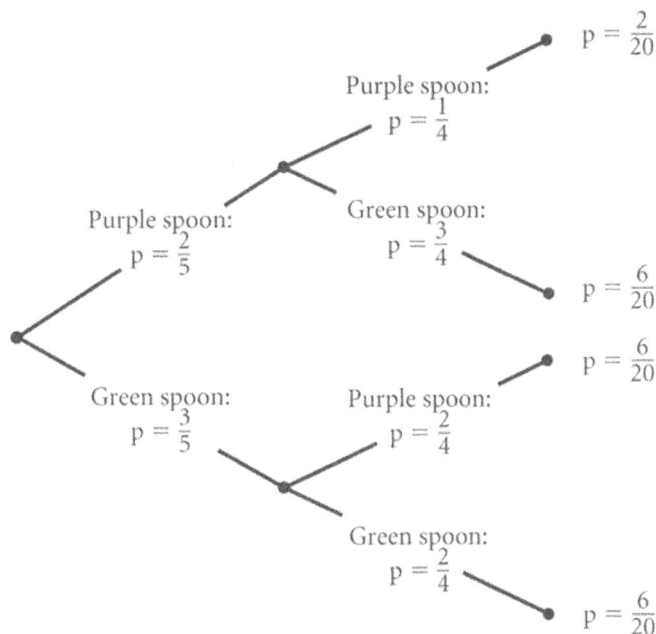

Purple spoon:
$p = \frac{1}{4}$
$p = \frac{2}{20}$

Purple spoon:
$p = \frac{2}{5}$

Green spoon:
$p = \frac{3}{4}$
$p = \frac{6}{20}$

$p = \frac{6}{20}$

Green spoon:
$p = \frac{3}{5}$

Purple spoon:
$p = \frac{2}{4}$

Green spoon:
$p = \frac{2}{4}$
$p = \frac{6}{20}$

Note: Unsimplified fractions are shown here for individual branches because they reflect the reasoning more clearly. Using unsimplified fractions for the final probabilities also more clearly represents the products of the fractions.

The principle of multiplying probabilities along a path of the diagram is an important idea, so you might reinforce it by going through the particular scenario using a "large number of cases" approach. To do this, have students consider what would happen in 100 trials of the spoon scenario. Students should see, for instance, that Scott and Letitia would expect their first spoon to be purple in about $\frac{2}{5}$ of the cases (which is 40 times) and that in 30 of these cases ($\frac{3}{4}$ of them), the second spoon would be green. Altogether, the "purple, green" sequence would occur 30 times out of 100, which corresponds to the fraction $\frac{30}{100}$, or $\frac{3}{10}$, which is equal to the product of the fractions $\frac{2}{5}$ and $\frac{3}{4}$.

You may find it helpful to think about two purposes for trees: one is to get a list of possible sequences; the other is to find the probability of each sequence. Students will be using them in both ways.

Part II: Allowance Choices

You can use the discussion of Part II to reinforce the ideas just discussed for using tree diagrams. If students think of the two bills (in Choice 2) as being pulled out of the bag one at a time, the tree diagram might look like this.

Second bill is $1:
$p = \frac{3}{4}$

$p = \frac{3}{5}$

First bill is $1:
$p = \frac{4}{5}$

Second bill is $5:
$p = \frac{1}{4}$

$p = \frac{1}{5}$

First bill is $5:
$p = \frac{1}{5}$

Second bill is $1:
$p = 1$

$p = \frac{1}{5}$

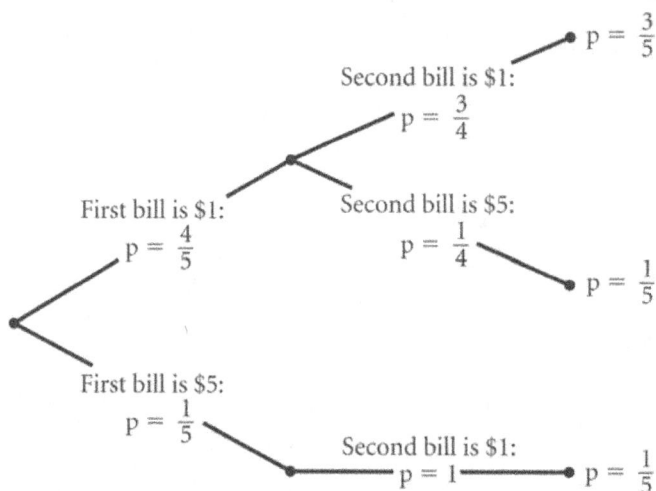

The uppermost path leads to a result of $2, and this result has a probability of $\frac{4}{5} \cdot \frac{3}{4}$, which equals $\frac{12}{20}$ (shown as simplified to $\frac{3}{5}$). The probability of getting $6 is $\left(\frac{4}{5} \cdot \frac{1}{4}\right) + \left(\frac{1}{5} \cdot 1\right)$, which simplifies to $\frac{2}{5}$.

Finding the Expected Value

The specific question in Part II is, "Which choice is better?" Students could argue that because $2 is more likely than $6 (as just shown), the average under Choice 2 will be less than $4 (which is the result of Choice 1). If they give this response, acknowledge that they are correct, but also ask what the exact average is. That is, have students compute the expected value.

As needed, review the idea of using a sample number of trials, such as 100. Of these, one would expect about $\frac{3}{5}$ of them, or about 60, to result in a $2 payment, and about $\frac{2}{5}$ of them, or about 40, to result in a $6 payment. That gives a total of (60 · $2) + (40 · $6), which is $360, for the 100 trials, or an average of $3.60. Thus, the expected value for Choice 2 is $3.60.

Ask, **What does $3.60 mean in terms of the problem?** Students should realize this means that in the long run, Choice 2 will not pay off as well as Choice 1. (Of course, this doesn't necessarily mean that Scott and Letitia will pick Choice 1. They may prefer to gamble.)

Key Questions

What does *expected value* mean?

How might you label this tree diagram to show probabilities?

What is the probability that the outcome will be the path P_1G_1?

How is this probability related to the probabilities for P_1 and G_1?

Why do you multiply these two probabilities?

How might you combine branches to simplify the diagram?

Doesn't this show that the probability of getting a purple spoon and then a green one is $\frac{1}{4}$?

What is the probability for each branch and subbranch?

What is the probability of the path "purple, green"?

What does $3.60 mean in terms of the problem?

Baseball Probabilities

Intent

This activity returns to the context of baseball to continue the review of probability.

Mathematics

Baseball Probabilities describes a situation with a baseball player whose batting average is known and consistent. The questions ask for the probabilities of various combinations of hits and failures to hit. This activity furthers the discussion of area and tree diagrams, focusing on ways to explain the multiplication of probabilities. Also reinforced is the fact that the sum of the probabilities across all possible distinct outcomes is 1.

Progression

Students work on this activity individually and then discuss their work as a class.

We recommend that you discuss *Choosing for Chores* before discussing *Baseball Probabilities*, because the use of tree diagrams to find probabilities is easier to understand in connection with that activity.

Approximate Time

30 minutes for activity (at home or in class)

20 minutes for discussion

Classroom Organization

Individuals, followed by whole-class discussion

Doing the Activity

Students should be able to complete this activity independently.

Discussing and Debriefing the Activity

Use the discussion to strengthen students' understanding of the principle of multiplying probabilities for sequences of events. Having students share a variety of approaches will be beneficial.

Question 1a

Question 1a is the simplest of the questions, so students should be able to justify their reasoning. For instance, if students say that the answer to Question 1a is $\frac{1}{9}$

because $\frac{1}{3} \cdot \frac{1}{3} = \frac{1}{9}$, have them explain why they want to multiply $\frac{1}{3} \cdot \frac{1}{3}$. Here are several possible explanations:

A fraction of a fraction: One approach is to say something like:

> *Willie gets a hit on his first time at bat $\frac{1}{3}$ of the time. He gets a hit on his second time at bat $\frac{1}{3}$ of that original $\frac{1}{3}$, and $\frac{1}{3}$ of $\frac{1}{3}$ is $\frac{1}{9}$, so he gets two hits $\frac{1}{9}$ of the time.*

Counting cases: For some students, the fraction-of-a-fraction reasoning can be clarified by working with a convenient specific number of cases. For instance, if Willie comes up to bat 36 times altogether, on average he would get a hit on his first time at bat 12 times (because $\frac{1}{3}$ of 36 is 12). Of those 12 occasions, he would get a hit on his second time at bat 4 times (because $\frac{1}{3}$ of 12 is 4). Thus, on average he gets a hit both times in 4 occasions out of 36, which is $\frac{1}{9}$ of the total.

An area diagram: An area diagram for Question 1a might have two stages:

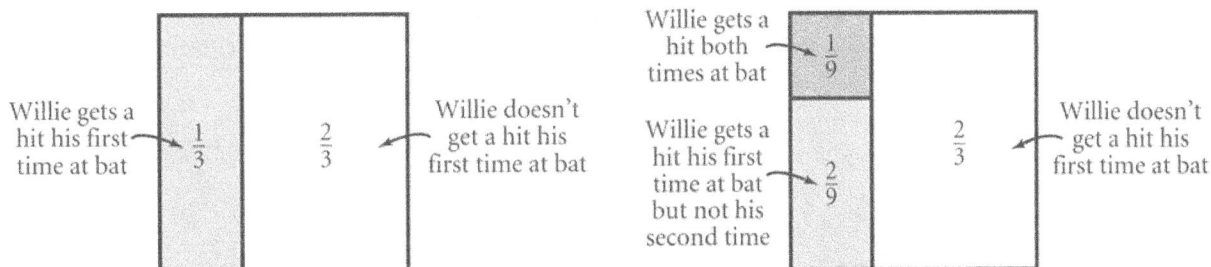

Note that it isn't necessary to subdivide the right side of the diagram for Question 1a, because for that question, it doesn't matter what happens to Willie his second time up if he doesn't get a hit the first time.

The main goal of this discussion is for students to see the connection between the arithmetic of the first two approaches and the areas in the diagram. Students should see that for Question 1a, they are interested in the area of the upper-left corner, and that this area can be found as the product of length and width. Because Willie's probability for each time at bat is $\frac{1}{3}$, both the length and width of this section of the diagram are $\frac{1}{3}$, so the area is $\frac{1}{3} \cdot \frac{1}{3}$, which is $\frac{1}{9}$.

This problem provides a good opportunity to review the idea that in a phrase like "$\frac{1}{3}$ of the times at bat," the word of can be interpreted arithmetically to mean *times*. It may help to do this in the context of an analysis like the "36 cases" approach as well as in the more abstract discussion of fractions. For instance, help students see that one can find $\frac{1}{3}$ of 36 using the arithmetic expression $\frac{1}{3} \cdot 36$ and that, similarly, one can find $\frac{1}{3}$ of $\frac{1}{3}$ using the arithmetic expression $\frac{1}{3} \cdot \frac{1}{3}$. (Students are likely to find $\frac{1}{3}$ of 36 by dividing 36 by 3, so it may be necessary to point out that division by 3 is the same as multiplication by $\frac{1}{3}$.)

The connection between *of* and *times* can also be explained in terms of the area diagram. The area representing hits on both at-bats is $\frac{1}{3}$ of $\frac{1}{3}$ of the total area, and this area can be found from its length and width as the product $\frac{1}{3} \cdot \frac{1}{3}$.

Question 1b

Question 1b is similar to Question 1a, except that the relevant fraction is $\frac{2}{3}$ instead of $\frac{1}{3}$. As before, have students give an explanation using an area diagram. They can continue with the diagram they used in Question 1a, except that the right-hand section of the diagram now needs to be subdivided. The diagram might look like this:

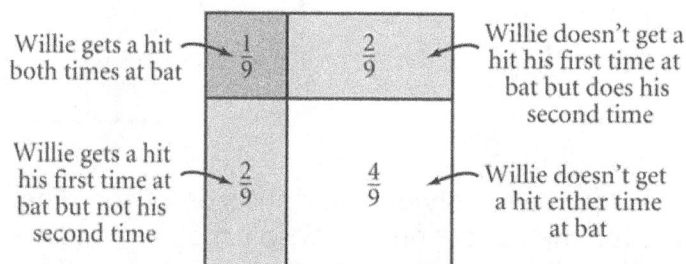

For Question 1b, students are interested in the lower-right section of the diagram, which is a rectangle whose length and width are $\frac{2}{3}$. Thus, this area is $\frac{2}{3} \cdot \frac{2}{3}$, which is $\frac{4}{9}$.

Question 1b using a Tree Diagram

Question 1b is sufficiently complex to make a tree diagram worthwhile. The diagram might look like this.

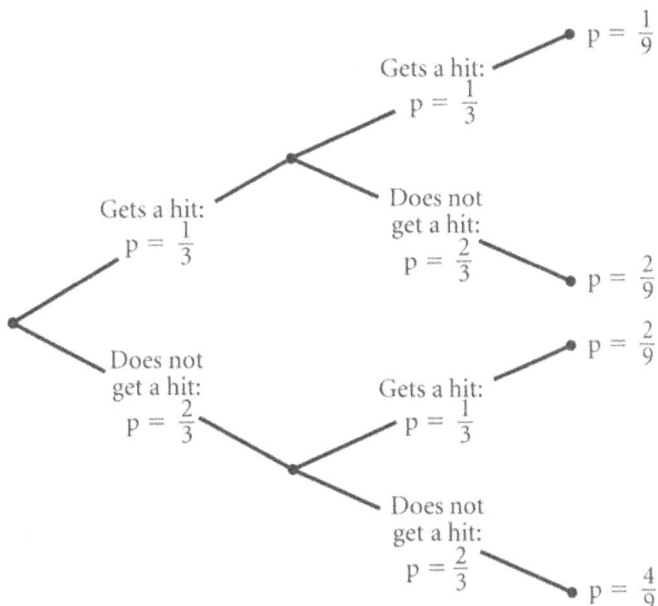

For Question 1b, students are concerned with the "bottom" path. Be sure they connect the fact that the probability for this path is the product of the probabilities for the two segments along the path with the multiplication they did to find areas in the area diagram.

Be sure students notice that this diagram has four paths altogether and that these paths (and their probabilities) correspond to the four sections of the area diagram.

Question 1c

Once students have a fairly clear idea about the probabilities in Questions 1a and 1b, move on to Question 1c. Although students can find this probability by combining two parts of the area diagram (or two paths of the tree diagram), the question specifically asks them to use their answers to Questions 1a and 1b.

Elicit the idea that the answer to Question 1c can be found by subtracting the sum of the answers to Questions 1a and 1b from 1. Have students justify why this makes sense. (Students will be using a similar idea in a more complex setting later in the unit, so it's worth noting here, even though it's not needed.)

Question 2

Questions 2a and 2b are similar to Questions 1a and 1b, but the extra time at bat makes the area diagram more complicated. For instance, a three-stage diagram focusing on Question 2a might look like this:

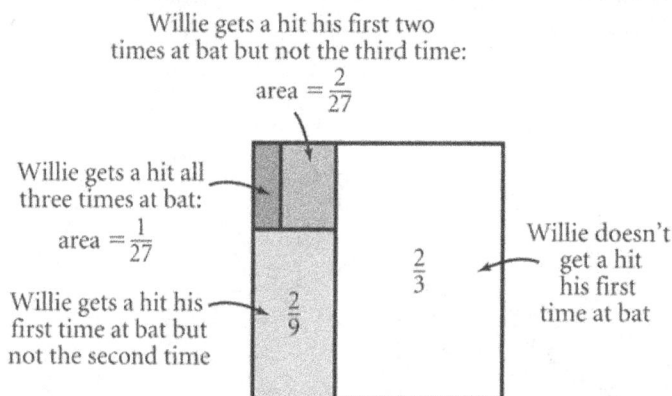

Willie gets a hit his first two times at bat but not the third time: area = $\frac{2}{27}$

Willie gets a hit all three times at bat: area = $\frac{1}{27}$

Willie gets a hit his first time at bat but not the second time

Willie doesn't get a hit his first time at bat

$\frac{2}{3}$

$\frac{2}{9}$

As before, there are various ways to think about the area of the section representing Willie getting three hits (the upper-left corner). Here are some possible explanations:

- It is $\frac{1}{3}$ of the area previously marked as $\frac{1}{9}$, and $\frac{1}{3}$ of $\frac{1}{9}$ is $\frac{1}{27}$.

- Its height is $\frac{1}{3}$ and its width is $\frac{1}{9}$, so its area is $\frac{1}{3} \cdot \frac{1}{9}$.

- The entire figure can be subdivided into 27 pieces of this size.

You can also have students give a "large number of cases" analysis, such as those described for Questions 1a and 1b.

Question 2a using a Tree Diagram

Students will probably appreciate the fact that as the situation gets more complex, an area diagram becomes more cumbersome and a tree diagram becomes more appealing. In particular, one needs to draw only a portion of the tree diagram, leaving out the subbranches that no longer matter. For instance, a tree diagram for the sequence "hit, no hit, no hit" might look like this, showing that the probability for this sequence is $\frac{1}{3} \cdot \frac{2}{3} \cdot \frac{2}{3}$, which equals $\frac{4}{27}$:

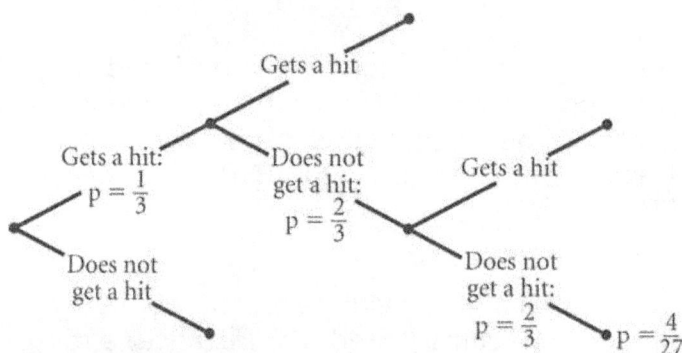

Gets a hit

Gets a hit: p = $\frac{1}{3}$

Does not get a hit: p = $\frac{2}{3}$

Gets a hit

Does not get a hit

Does not get a hit: p = $\frac{2}{3}$

p = $\frac{4}{27}$

Question 2b

You may want to have students make both an area diagram and a tree diagram for Question 2b. This will give them practice in the two methods and bring out how elements of the two approaches correspond. They should see that the probability here is simply $\frac{2}{3} \cdot \frac{2}{3} \cdot \frac{2}{3}$, which equals $\frac{8}{27}$.

Question 2c

For Question 2c, be sure students understand that they didn't need to find the probabilities of one hit and two hits separately, but were asked only for the combined probability. As with Question 1c, this problem focuses on the principle that the probabilities all add up to 1, so students can get an answer using the expression $1 - \frac{1}{27} - \frac{8}{27}$, which equals $\frac{18}{27}$. (You needn't simplify this fraction.)

Key Question

Can you draw an area model or a tree diagram that shows the possible sequence of events?

Possible Outcomes

Intent

This activity resumes the focus on the central unit problem.

Mathematics

Possible Outcomes provides practice with systematically listing the various cases in a complex situation: students list all of the possible records that each of the two teams in the unit problem might have for the remaining seven games, and then enumerate the possible combinations of records for the two teams together. The activity also establishes the multiplication principle for the number of ways to pick one object from each of two sets.

Progression

Students work on the activity individually and then discuss their work in class.

Approximate Time

30 minutes for activity (at home or in class)

15 minutes for discussion

Classroom Organization

Individuals, followed by whole-class discussion

Materials

Possible Outcomes blackline master (1 copy per student)

Doing the Activity

Students should be able to complete this activity independently.

Discussing and Debriefing the Activity

Students should see that there are eight possibilities for each team, because the number of their wins can be any integer from 0 through 7. (Some students may get seven possibilities by forgetting 0; clarify this situation as needed.) They should also see that having eight possibilities for each team means that there are 64 possible combinations.

Have a volunteer share his or her display for Question 3b. If students have alternative ways to present the combinations, let them share these ideas. A display like the one below is a particularly effective way to show all combinations. The individual boxes within this chart can be referred to as *cells*. In this display, a "G" in a cell indicates that the Good Guys win the pennant if that combination occurs, a "B" means the Bad Guys win, and a "T" means a tie (that is, the two teams end up with identical records for the season). Provide a copy of this chart (the *Possible Outcomes* blackline master) to each student; it has space for students to write in the probability of each combination as it's found throughout the unit. You may also want to post a master copy of the chart in the classroom and fill in probabilities as students find them.

Good Guys' record for the final seven games									
		7–0	6–1	5–2	4–3	3–4	2–5	1–6	0–7
Bad	7–0	G	G	G	T	B	B	B	B
Guys'	6–1	G	G	G	G	T	B	B	B
record	5–2	G	G	G	G	G	T	B	B
for	4–3	G	G	G	G	G	G	T	B
the	3–4	G	G	G	G	G	G	G	T
final	2–5	G	G	G	G	G	G	G	G
seven	1–6	G	G	G	G	G	G	G	G
games	0–7	G	G	G	G	G	G	G	G

To orient students to what the cells in the chart represent, ask, **Which cell represents both the Good Guys and the Bad Guys winning all their remaining games?**

Point out that 49 of the 64 cells show the Good Guys as the winner, and ask, **Doesn't this chart show that the Good Guys' probability of winning is $\frac{49}{64}$? Why not?** Students will probably see that it does not, but ask for an explanation of how they know this.

If students do not see this or do not have an explanation, ask, **Are the Good Guys as likely to win all their games as to lose them all?** They should see that "all wins" is more likely than "all losses" and that this is also true for the Bad Guys. Thus, the case in the upper-left corner of the table is more likely than the case in the lower-right corner. (In *How Likely Is All Wins?*, they will actually find the probabilities of these results.)

Ask, **What records do you think are most likely?** (This is similar to Question 2 of *Race for the Pennant!*) Many students will probably see intuitively that the most likely outcome for each team is a record of four wins and three losses, because this gives a percentage as close as possible to the teams' current winning rates of .62 and .6.

Tell students that one of their main tasks in the unit will be to find the probability for each of the cells in this chart.

The Chart is not an Area Diagram

You might ask students whether this chart is an area diagram—it's not, because the individual cells are not equally likely. If you like, let students speculate on how the diagram would have to be modified for the area of each cell to represent the probability of the corresponding outcome.

Question 3a and the Multiplication Principle for Counting

Question 3a illustrates the general principle that to find the number of combinations when taking one object from each of two sets, one simply multiplies the number of objects in each set. The two-dimensional layout of the chart provides a nice way to see this principle. You may want to call this the *multiplication principle* so you can refer to it later and distinguish it from other counting principles.

This is a much simpler counting principle than the concepts of permutations and combinations that students will encounter later in the unit. The multiplication principle generalizes to any number of sets: If set A_1 has n_1 elements, set A_2 has n_2 elements, and so on through set A_t with n_t elements, then the number of ways of picking one element out of each set is the product $n_1 \cdot n_2 \cdot \ldots \cdot n_t$. *(Be careful that students do not confuse this multiplication principle for counting with the principle of multiplying probabilities for sequences of events, as used in finding the probability for a path in a tree diagram. Though the two principles are related, they are quite distinct.)*

Key Questions

Which cell represents both the Good Guys and the Bad Guys winning all their remaining games?

Doesn't this chart show that the Good Guys' probability of winning is $\frac{49}{64}$? Why not?

Are the Good Guys as likely to win all their games as to lose them all?

What records do you think are most likely?

Supplemental Activities

Putting Things Together (reinforcement) gives students some additional experience with the multiplication principle. This activity would be especially useful in advance of *Who's on First?*, as the multiplication principle reappears in Question 4 of that activity.

How Likely Is "All Wins"?

Intent

In this activity, students find the probability for multiple events occurring.

Mathematics

How Likely Is All Wins? asks students to find the probabilities for each team winning all of their remaining games and then the probability of both teams finishing that way. The discussion emphasizes the use of tree diagrams and area models, and introduces the notion that the probabilities could be calculated without using either method.

Progression

Students work on the activity in groups or individually and then discuss their work as a class.

Approximate Time

30 to 35 minutes for activity

15 to 20 minutes for discussion

Classroom Organization

Groups, followed by whole-class discussion

Doing the Activity

Tell students that they will continue their investigation of the central unit problem by finding the probability that the Good Guys and the Bad Guys will each win all seven of their remaining games. Ask students which cell of the chart this case corresponds to.

If students need a hint on Question 3, point out that the overall chart can be thought of as a not-to-scale area diagram and that Question 3 asks them to find the area of one of the rectangles in this diagram. Questions 1 and 2 give the length and width of that rectangle.

Discussing and Debriefing the Activity

Have a student report on Question 1, giving the group's reasoning as well as the numerical answer. Ask if other groups have alternate explanations, trying to elicit both an area diagram and a tree diagram. Reasoning similar to that used in several recent activities should show that the answer is $.62^7$, which is approximately .0352.

Students may have difficulty providing diagrams for this problem because they have to use so many stages of branching or divide the area so many times. The upcoming activity *Diagrams, Baseball, and Losing 'em All* will invite discussion about the relative advantages of area and tree diagrams and about the option of not using any diagram.

Questions 2 and 3

Once you have discussed Question 1 fully, you can probably simply get an answer for Question 2, skipping the discussion. The numerical value of $.6^7$ is approximately .0280.

Question 3 involves putting these two results together to find the probability that both events will happen. This problem differs slightly from most of the recent problems because the results for the two teams do not form a sequence of events. This makes a tree diagram less natural as a model, and an area diagram perhaps more appealing.

You can elicit that the probabilities would be the same if the Good Guys played all of their games before the Bad Guys played any of theirs. This may help students see that they should multiply the two probabilities in Question 3, as they have in other problems. This gives an overall probability of about $.0352 \cdot .0280$, which is approximately .0010, for the outcome of both teams winning all their games.

Ask, **Which cell(s) in the chart from Possible Outcomes does this probability apply to?** There is only one—have students identify this cell and enter the probability. That's one down and 63 to go!

Students may point out that if the Good Guys win all their games, then it doesn't matter what the Bad Guys do. Therefore, as far as finding the Good Guys' chances of winning the pennant, students don't necessarily need to find the probability for every cell in the chart. However, we will find all the individual probabilities as the unit progresses for the sake of completeness.

Key Question

Which cell(s) in the chart from Possible Outcomes does this probability apply to?

Go for the Gold!

Intent

In this activity, students continue to analyze probabilities using area diagrams, tree diagrams, and expected value.

Mathematics

Go for the Gold! builds on the theme introduced in *Choosing for Chores* to explore the probability of a sequence of independent events. The activity describes a game that you must pay to play. Students are asked to find the probability that the winning sequence of events will occur, using both a tree diagram and an area model. They are then asked whether they should accept an invitation to play the game—a question best evaluated using expected value.

Progression

Students work on the activity individually and then discuss their work as a class.

Approximate Time

30 minutes for activity (at home or in class)

25 to 30 minutes for discussion

Classroom Organization

Individuals, followed by whole-class discussion

Doing the Activity

Students should be able to complete this activity independently.

Discussing and Debriefing the Activity

Ask groups to come to a consensus on the answers and on justifications based on area diagrams and tree diagrams. Then ask each group to make a presentation.

An area diagram for Question 1a might proceed in two stages, with the first stage looking like this:

First cube is white:
$p = \frac{2}{3}$

$\frac{2}{3}$

First cube is gold:
$p = \frac{1}{3}$

$\frac{1}{3}$

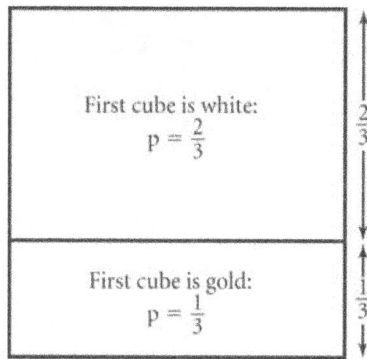

In the second stage, the bottom portion would be subdivided to show the color of the second cube. (There is no second cube if the first cube is white.) The shaded portion represents winning the game.

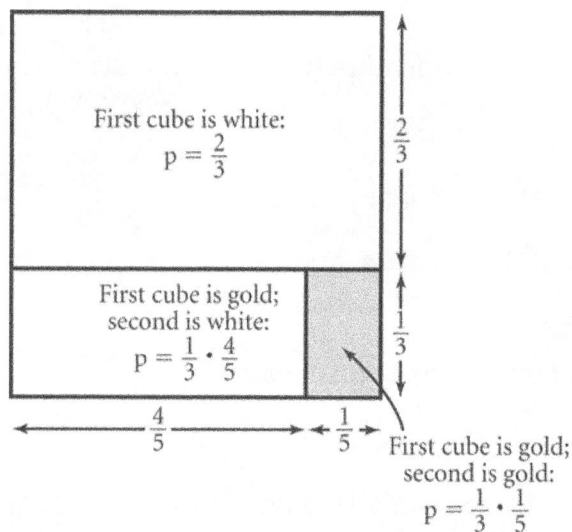

First cube is white:
$p = \frac{2}{3}$

$\frac{2}{3}$

First cube is gold;
second is white:
$p = \frac{1}{3} \cdot \frac{4}{5}$

$\frac{1}{3}$

$\frac{4}{5}$

$\frac{1}{5}$

First cube is gold;
second is gold:
$p = \frac{1}{3} \cdot \frac{1}{5}$

A tree diagram for the situation might look like this:

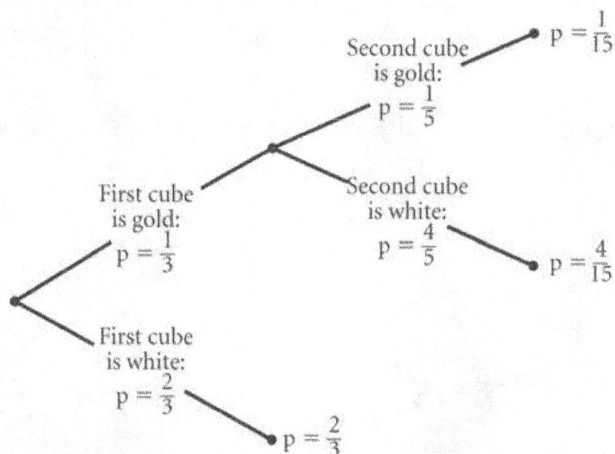

Second cube
is gold:
$p = \frac{1}{5}$

$p = \frac{1}{15}$

First cube
is gold:
$p = \frac{1}{3}$

Second cube
is white:
$p = \frac{4}{5}$

$p = \frac{4}{15}$

First cube
is white:
$p = \frac{2}{3}$

$p = \frac{2}{3}$

The area and tree diagrams show that the probability of winning is $\frac{1}{3} \cdot \frac{1}{5}$, which equals $\frac{1}{15}$.

Note: Some students may prefer a tree diagram in which each cube gets its own branch. They might also want to show a second cube being selected even if the first is white. Such a diagram would have 15 equally likely paths and might look like this:

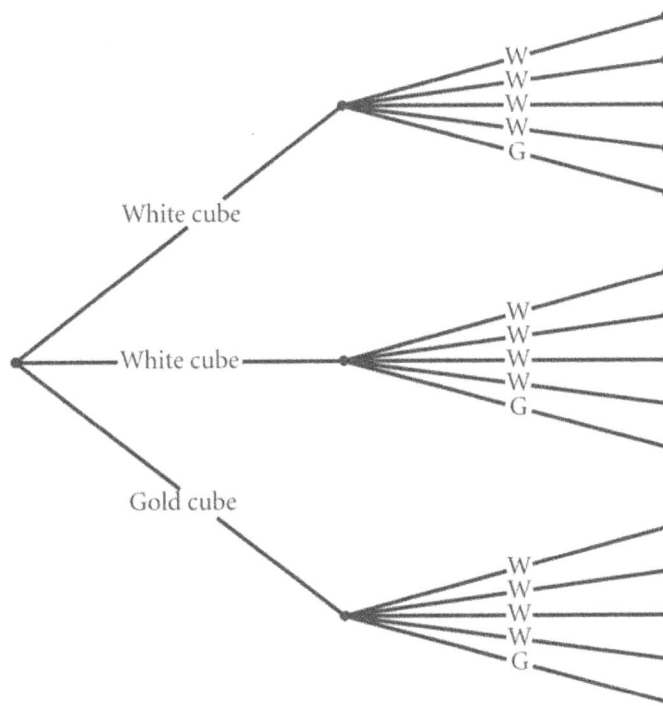

(Notice that if a tree diagram has equally likely paths, then it is less important to label the branches with their individual probabilities.)

After students have presented solution methods using both area diagrams and tree diagrams, have a volunteer describe how the two types of diagrams correspond. For instance, the second area diagram has three sections, and the first tree diagram has three complete paths (or three final nodes), and students can match them up.

Ask, **Can you explain the probabilities using a "large number of cases" approach?** For instance, using 30 games, students should see that they would get the first gold cube 10 times, and then, of those 10, they would get the second gold cube 2 times. Thus, they would win 2 times out of 30, which gives a probability of $\frac{1}{15}$.

Question 2

Once students are clear about why the probability of winning $1,000 is $\frac{1}{15}$, have someone answer Question 2.

Students might continue with the 30-game analysis, pointing out that they would win $1,000 twice, for a gain of $2,000, while spending $100 for each of their 30 games, for a loss of $3,000. Thus, the 30 games would result in a net loss of $1,000, so players would lose money in the long run.

You might ask students what the actual expected value is, even though the problem didn't ask for that. They should see that a net loss of $1,000 for 30 games means an expected value per game of –$33.33.

Students may legitimately disagree about whether to play or not. Clearly, playing represents a gamble. This is a good opportunity to discuss the difference between the expected value, which represents the average gain or loss per trial over the long run, and the outcome of a single trial, whose possibilities often differ from the expected value.

Key Question

Can you explain the probabilities using a "large number of cases" approach?

Supplemental Activity

Ring the Bells! (reinforcement) makes a good follow-up to *Go for the Gold!*

Diagrams, Baseball, and Losing 'em All

Intent

In this activity, students compare the use of area diagrams and tree diagrams in calculating probabilities.

Mathematics

This activity primarily gives students an opportunity to reflect on their work so far. In Question 3, students calculate the probability of one more case from the chart of possible outcomes.

Progression

Students work on this activity individually and then discuss their work as a class.

Approximate Time

30 minutes for activity (at home or in class)

10 minutes for discussion

Classroom Organization

Individuals, followed by whole-class discussion

Doing the Activity

Students should be able to complete this activity independently.

Discussing and Debriefing the Activity

You may want to skip discussion of Questions 1 and 2 and simply collect and read students' work on those questions to get a sense of their thinking on the issues.

Have a presentation on Question 3, and ask, Why do you multiply probabilities in Question 3? (A tree or area diagram will probably be helpful.) Then have students enter that result in the appropriate place in their charts. They should see that all losses for both teams has a probability of $.38^7 \cdot .4^7$, which is approximately .000002.

Bring out how different the probabilities are for the two cells that students have dealt with so far. (They found yesterday that the probability for all wins for both teams is approximately .0010.)

Key Question

Why do you multiply probabilities in Question 3?

The Birthday Problem

Intent

In the activities in *The Birthday Problem*, students explore a classic problem as they come to understand how to calculate the probability of a sequence of independent events.

Mathematics

Most of the activities in this section are variations of the "birthday problem:" How many people need to be in a room to have a probability greater than $\frac{1}{2}$ that at least two of them have the same birthday? Students see through these activities that the product of the probabilities of several independent events will give the probability of a sequence of those events occurring.

In pursuit of solutions to several variations of this problem, students will make repeated use of the **pigeonhole principle**. They will also observe that it is often easier to find the probability of a sequence of events occurring by calculating the probability of the sequence *not* occurring, and then subtracting that probability from 1.

Progression

The first two activities in this section, *POW 11: Let's Make a Deal* and *Simulate a Deal,* are not related to the birthday problem, but introduce a new POW. *Six for the Defense* asks students to answer a series of questions based on a tree diagram they have drawn of the situation. *Day-of-the-Week Matches* and *"Day-of-the-Week Matches" Continued* ease students into the birthday problem by giving them a variation that involves fewer possible outcomes—looking for individuals born on the same day of the week rather than the same day of the year. *Monthly Matches* continues the preparation for *The Real Birthday Problem*.

Students will not be discussing the baseball pennant problem in this cluster of activities. Throughout the rest of this unit, they'll alternate between focusing on the pennant problem, and focusing on other problems that may seem like digressions, but will in fact be solidifying students' understanding of tree diagrams and probability, and how to find probabilities for other cells in their chart.

POW 11: Let's Make a Deal

Simulate a Deal

Day-of-the-Week Matches

 Day-of-the-Week Matches, Continued

Monthly Matches

The Real Birthday Problem

Six for the Defense

POW 11: Let's Make a Deal

Intent

In this activity, students use probability to evaluate strategies for a game.

Mathematics

This POW describes a game show involving prizes behind three doors and asks students to evaluate the probability of success with each of two possible strategies. In their write-ups and presentations, students focus on explaining the probabilities.

Calculating the probabilities in this situation is particularly interesting because they are difficult to assess intuitively.

Progression

Students will gain some insight into this POW through a simulation in the subsequent activity, *Simulate a Deal*.

Approximate Time

5 minutes for introduction

1 to 3 hours for activity (at home)

15 to 20 minutes for presentations and discussion

Classroom Organization

Individuals, followed by several presentations and whole-class discussion

Doing the Activity

Have students read the POW, and give them an opportunity to ask questions about how the game is played. Tell students that in *Simulate a Deal* they will do a simulation of the two strategies and discuss their results. (This simulation will probably reveal the answer to the "big question" of the POW, but students will still have to determine the specific probabilities and explain them.)

Give students about a week to complete this POW.

On the day before the POW is due, choose three students to make presentations on the following day, and give them overhead transparencies and pens to take home to use for preparing those presentations.

Discussing and Debriefing the Activity

Ask three selected students to do their presentations. From the simulations in *Simulate a Deal,* they should have experienced that by staying with their guess they win about $\frac{1}{3}$ of the time, and by switching, they win about $\frac{2}{3}$ the time. The focus of the discussion should therefore be on why the two strategies have these probabilities of success. Explanations of this vary widely, from elaborate tree diagrams to very concise arguments.

There is a nice discussion of this problem in "Monty's Dilemma: Should You Stick or Switch?" by Shaughnessy and Dick, in NCTM's *Mathematics Teacher* (April 1991, pp. 252–256). The article suggests that the problem might be easier to understand in this more extreme form:

> *Suppose there are 100 doors altogether, and that after you guess, the host opens all the doors except yours and one other. Should you stay or switch?*

In this situation, students should be able to see that it makes sense to switch.

Simulate a Deal

Intent

In this activity, students use a simulation to evaluate strategies.

Mathematics

While the answer to the dilemma posed in the POW is difficult to arrive at intuitively, this simulation will convince most students of which strategy is best. After the class briefly discusses what makes a good simulation, students work in their groups to design and carry out the experiment. Afterwards, the class compiles a list of their simulation results.

Progression

Simulate a Deal asks students to devise and carry out a simulation to estimate the probabilities associated with *POW 11: Let's Make a Deal.*

Approximate Time

10 to 20 minutes for activity

5 to 10 minutes for discussion

Classroom Organization

Pairs, followed by whole-class discussion

Materials

A variety of materials that groups are likely to request for conducting a simulation, such as colored cubes, coins, paper bags

Doing the Activity

This activity is intended to point students in the right direction on their POW. After the discussion of the activity, they should be convinced that the best strategy for *POW 11: Let's Make a Deal* is always to switch. But they will likely not understand why this strategy is best. Explaining why will be the main task in the POW.

Discuss simulations before students begin the activity. You might first ask, When else have you used simulations? to elicit examples of simulations students have seen in other contexts.

What's a Good Simulation?

Ask, What makes a good simulation? Be sure that students note that randomness is essential in most simulations that involve probability. Therefore, a good simulation for this situation would be one in which the student playing the host chooses the winning door at random and the student playing the contestant makes the initial choice of door at random.

If students need help setting up their simulations, you can propose this approach:

1. One student (the host) decides where the car is (perhaps writing this on a piece of paper).
2. The other student (the contestant) chooses a door (stating the choice aloud).
3. The host "opens" a specific door that does not have the car. (The host may or may not have a choice about which door to open.)
4. The contestant chooses the remaining closed door (in Question 1) or announces the decision to stay with the door originally chosen (in Question 2).
5. The host reveals which door actually has the car behind it.

Discussing and Debriefing the Activity

Have one or two teams describe exactly how they carried out their simulations, and then have all teams share their results. Compile a class list of the number of wins and losses for each strategy. This should convince most students that switching is a better strategy than staying.

Point out to students that they will describe their work on this activity as part of their POW write-ups. Emphasize, however, that a key task of the POW is to explain *why* they (and the class) got the type of results they did.

Key Question

When else have you used simulations?

What makes a good simulation?

Supplemental Activities

Programming a Deal (extension) may be appealing to any programming fans in your class

Simulation Evaluation (extension) asks students to use the chi-square statistic to evaluate the reliability of their conclusion that switching is a better strategy than staying.

Day-of-the-Week Matches

Intent

In this activity, students solve a probability problem using the pigeonhole principle.

Mathematics

Day-of-the-Week Matches poses several questions regarding the probability of finding that individuals chosen at random were born on the same day of the week. Discussion of the first question introduces the **pigeonhole principle**. Discussion of the final question introduces the idea that it may be easier to find the probability that an event occurs by subtracting the probability that it does not occur from 1.

Progression

This assignment begins a multi-day digression that culminates in the activity *The Real Birthday Problem*. This work will strengthen students' ability to use tree diagrams and area models and improve their understanding of the probabilities for multistage events.

Approximate Time

30 minutes for activity (at home or in class)

35 to 50 minutes for discussion

Classroom Organization

Individuals, followed by whole-class discussion

Discussing and Debriefing the Activity

The assignment contains important ideas, so allow plenty of time for students to ask questions and share alternative ways of approaching the questions.

Question 1

Have a student report on Question 1. The presenter should be able to explain that with eight people, you can be sure that at least two people were born on the same day of the week.

Tell the class that this is an example of the **pigeonhole principle**, which occurs in many contexts in mathematics. (The name comes from the image of trying to put pigeons into pigeonholes. The principle states that if you have more pigeons than you have pigeonholes, you will need to put at least two pigeons in the same pigeonhole.)

Ask, **Have you used the pigeonhole principle before?**

Question 2

Two approaches likely will come up in the discussion of Question 2. One method is to think of one of the people as "first" and the other as "second." Using that approach, once the first person is chosen, the issue is simply whether the second person's birth day-of-the-week is the same as the first's. Because (by assumption) all seven days of the week are equally likely, students should see that the

probability of a match is simply $\frac{1}{7}$.

The second approach is to look at all *pairs* of days of the week for the two people. From this perspective, there are 49 cases to consider, and in exactly 7 of these cases, the days of the week are the same for the two people. From this perspective,

one might see the probability of a match as $\frac{7}{49}$ rather than as $\frac{1}{7}$ (but, of course,

the two fractions are equal).

If students think of the people as being chosen sequentially, then a tree diagram is a natural tool for describing the situation. If the two people are thought of as having "equal status," then an area diagram (with a 7 x 7 grid) may seem more natural. However, in preparation for Question 3, we suggest that you elicit tree diagrams for both approaches.

Ask for a volunteer to present Question 2, and then elicit other approaches. One simple tree diagram representing this situation looks like this:

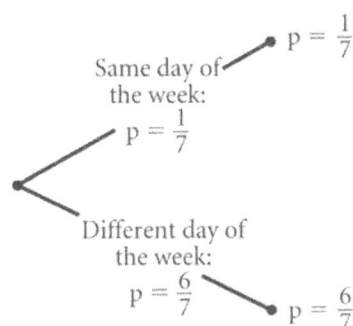

Alternatively, the tree diagram might have an initial branch for each day of the week, like this:

Second born Sunday: $p = \frac{1}{49}$

$p = \frac{1}{7}$

Second not born Sunday: $p = \frac{6}{7}$ → $p = \frac{6}{49}$

First born Sunday: $p = \frac{1}{7}$

Second born Monday: $p = \frac{1}{49}$

$p = \frac{1}{7}$

Second not born Monday: $p = \frac{6}{7}$ → $p = \frac{6}{49}$

First born Monday: $p = \frac{1}{7}$

Second born Tuesday: $p = \frac{1}{49}$

$p = \frac{1}{7}$

First born Tuesday: $p = \frac{1}{7}$

Second not born Tuesday: $p = \frac{6}{7}$ → $p = \frac{6}{49}$

etc.

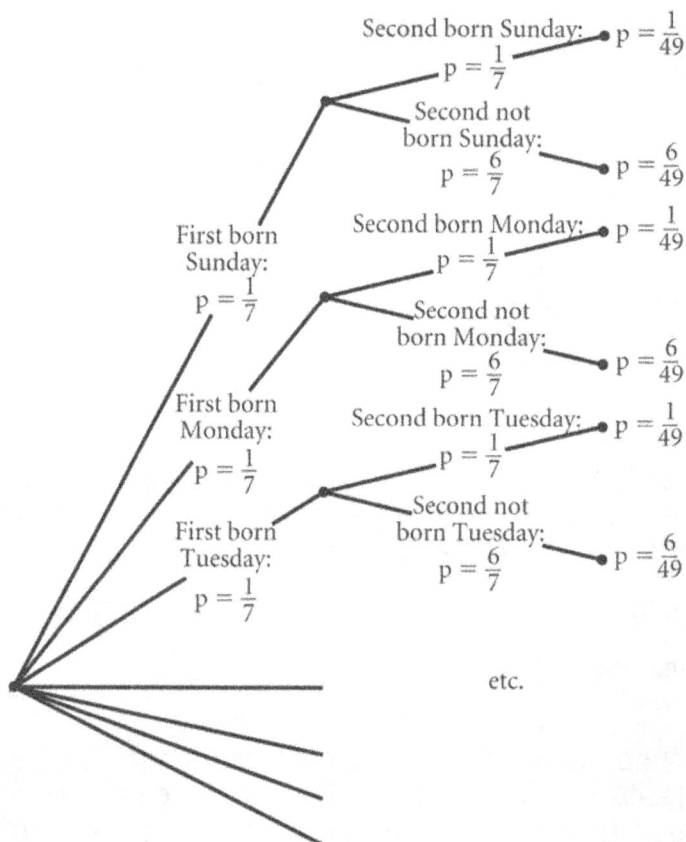

In using this second tree diagram, one needs to identify the paths that lead to matches (the upper subbranch for each initial branch) and add up the corresponding probabilities.

Students might also use a 7 x 7 grid to illustrate the possible combinations of birth days-of-the-week for two people, but that approach is limited beyond the case of two people. Encourage different approaches, and ask, **Why do these methods give the same answer?**

Post the theoretical result that the probability of a match is $\frac{1}{7}$ for comparison with an experimental approach following *"Day-of-the-Week Matches" Continued*.

Question 3

Question 3 involves a significantly more complex situation than Question 2. For this problem, it becomes more important to think of the people as being selected sequentially. One key element in the analysis is the realization that if the second person's birth day-of-the-week matches the first person's, then it doesn't matter on what day of the week the third person was born. But if the first two people were born on different days of the week, then there are two days of the week that the third person can match.

We suggest that you try to elicit both a tree diagram and an area diagram for Question 3. You may want to save both of these diagrams in case students need to build on them in the discussion of *"Day-of-the-Week Matches" Continued*.

A tree diagram might build on the first version shown for Question 2:

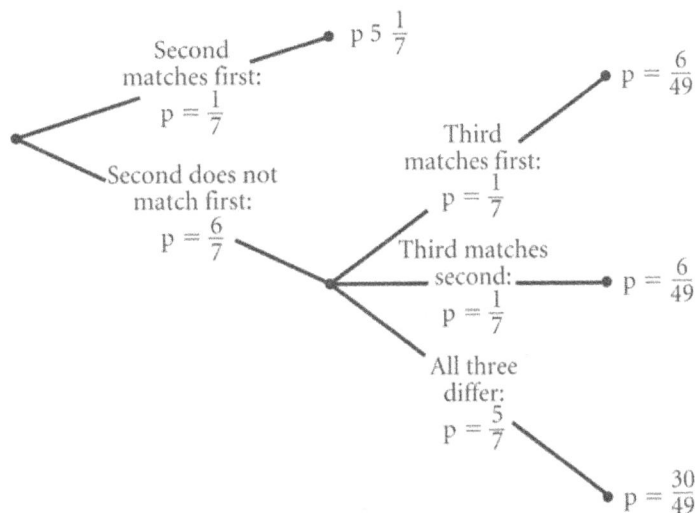

Second matches first: $p = \frac{1}{7}$ p 5 $\frac{1}{7}$

Second does not match first: $p = \frac{6}{7}$

Third matches first: $p = \frac{1}{7}$ $p = \frac{6}{49}$

Third matches second: $p = \frac{1}{7}$ $p = \frac{6}{49}$

All three differ: $p = \frac{5}{7}$ $p = \frac{30}{49}$

Thus, the probability of a match is $\frac{1}{7} + \frac{6}{49} + \frac{6}{49}$, which is $\frac{19}{49}$, or about 39%.

Some students may find an area diagram useful in explaining where the fractions $\frac{6}{49}$ and $\frac{30}{49}$ come from. As usual, the diagram will probably be done in two stages. The first stage might look like this:

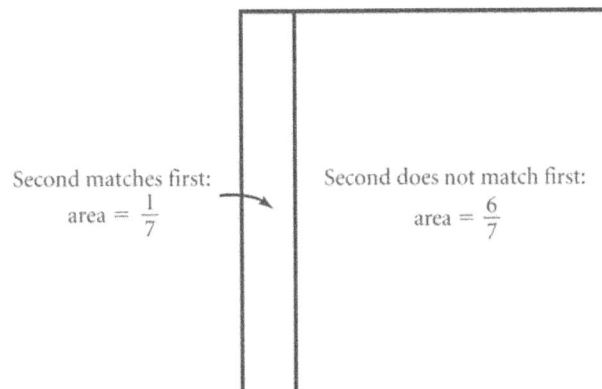

Second matches first: area = $\frac{1}{7}$ Second does not match first: area = $\frac{6}{7}$

In the second stage, the right side of the diagram is divided to indicate whether the third person matches either of the first two (and if so, whom that person matches).

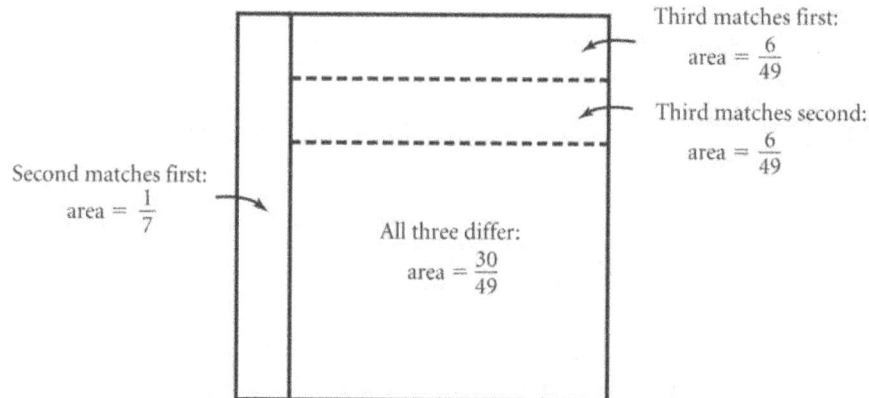

Ask, **Couldn't the third person match both of the first two? Where does this appear on the diagram?** Be sure students see that this possibility is included within the area labeled "Second matches first" and represents $\frac{1}{7}$ of that area.

This is important because it explains why the areas labeled "Third matches first" and "Third matches second" do not overlap. That is, these areas should more accurately be labeled "Second does not match first, and third matches first" and "Second does not match first, and third matches second."

This distinction may be clearer in the tree diagram, where the subbranches labeled "Third matches first" and "Third matches second" are seen coming from the branch "Second does not match first."

Finding the Probability of No Matches

For the follow-up assignment, *"Day-of-the-Week Matches" Continued*, and in preparation for *The Real Birthday Problem*, it's important for students to see how to find the probability of getting no matches without calculating the probabilities for all possible individual outcomes.

Ask, **How can you find the probability of getting no matches without analyzing every case?** Students might give an answer something like this:

The probability is $\frac{6}{7}$ that the second person will have a different birth day-of-

the-week from the first. If that happens, then there is a $\frac{5}{7}$ chance that the *third person's birth day-of-the-week will be different from both of theirs. That gives an overall probability of $\frac{6}{7} \cdot \frac{5}{7}$, which equals $\frac{30}{49}$.*

Some students might want to consider a "many cases" approach, perhaps using 49 trials. In $\frac{6}{7}$ of the cases (which is 42), the second person will have a different birth day-of-the-week from the first. Then, in $\frac{5}{7}$ of those 42 cases, the third person will have a different birth day-of-the-week from both the first person and the second person. This gives 30 non-match cases out of 49.

Then ask, **How can you use the probability of no matches to find the probability that there is at least one match?** Help students as needed to see that they can get this by subtracting $\frac{30}{49}$ (which they just found to be the probability of no matches) from 1, giving $\frac{19}{49}$. Ask them to check this by adding the probabilities for there being a match (from the area diagram or tree diagram). They will see that the separate cases that give a match have probabilities $\frac{1}{7}$, $\frac{6}{49}$, and $\frac{6}{49}$ and that these give a total that is equal to $1 - \frac{30}{49}$.

Point out that students have seen two fairly different ways to get the probability of having at least one match:

- They can find the probability of having no matches (simply $\frac{6}{7} \cdot \frac{5}{7}$, which equals $\frac{30}{49}$) and then subtract this result from 1.
- They can calculate the probabilities for all of the separate cases in which there is a match and add the results $\left(\frac{1}{7} + \frac{6}{49} + \frac{6}{49} \right)$.

Elicit the observation that the first approach is probably somewhat simpler, both conceptually and computationally, and suggest that students keep this idea in mind for future similar problems.

Key Questions

Have you used the pigeonhole principle before?

Why do these methods give the same answer?

Couldn't the third person match both of the first two? Where does this appear on the diagram?

How can you find the probability of getting no matches without analyzing every case?

How can you use the probability of no matches to find the probability that there is at least one match?

Day-of-the-Week Matches, Continued

Intent

In this activity, students use conditional probability, continuing the theme from *Day-of-the-Week Matches.*

Mathematics

In *Day-of-the-Week Matches, Continued,* students apply the principle that the probability of an event *not happening* is 1 minus the probability of the event *happening*. The activity asks what the minimum number of people is such that the probability of having at least two of them born on the same day of the week is greater than $\frac{1}{2}$. The discussion afterward demonstrates the associated probabilities using a tree diagram, an area diagram, and by multiplying probabilities without a diagram. If students know what day of the week they were born (from their work on *POW 10: Happy Birthday!),* they can see how closely your class matches the theoretical probabilities.

Progression

Students work on the activity individually and then discuss their results as a class.

Approximate Time

25 minutes for activity (at home or in class)

30 to 50 minutes for discussion

Classroom Organization

Individuals, followed by whole-class discussion

Doing the Activity

Students should be able to complete this activity independently.

Discussing and Debriefing the Activity

Students should find that for four people, the probability of having a match is $\frac{223}{343}$ (approximately 65 percent). Because this is more than $\frac{1}{2}$, and because students found in *Day-of-the-Week Matches* that the probability of a match for three people is less than $\frac{1}{2}$, the answer to the activity's question is "four people."

Have at least one student give a justification using a tree diagram, at least one explain it using an area diagram, and at least one explain it without a diagram. Following are some suggestions for these different approaches.

Using a Tree Diagram

A tree diagram for the problem might look like this.

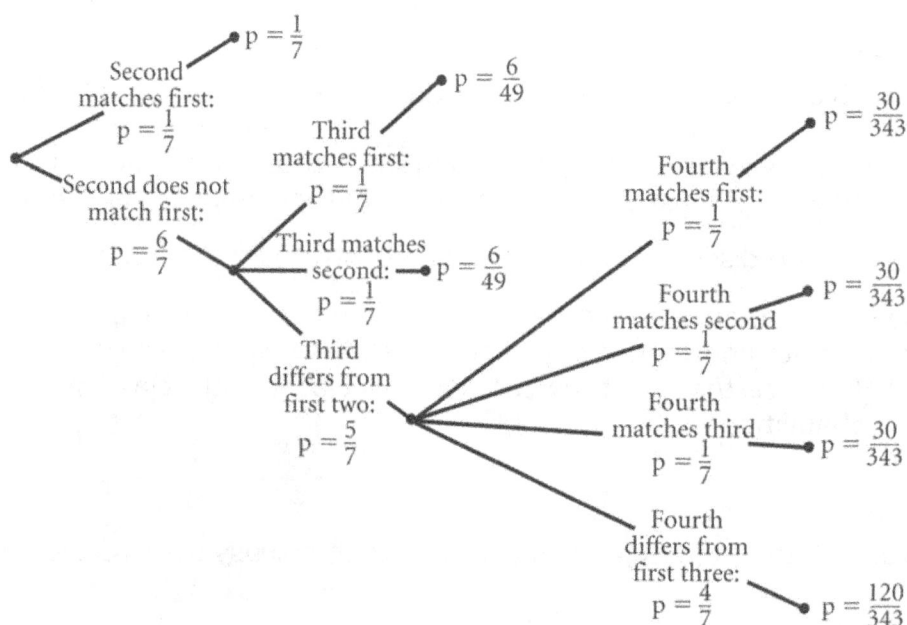

As usual, students should explain where the various probabilities are coming from. In particular, if they multiply the probabilities along the branches, ask, **Why does this multiplication process work?** (They may wish to use an area model to justify this aspect of their reasoning.)

Once students have the various end-node probabilities, they can either add up the fractions that represent matches or subtract the probability of "no match" from 1. (They will likely rather do the latter. In fact, to answer the question, they don't need to find the probabilities for each final node.)

Using an Area Diagram

If students need help in generating an area diagram for the four-person situation, you can begin with the diagram for the case of three people created in the discussion of *Day-of-the-Week Matches*:

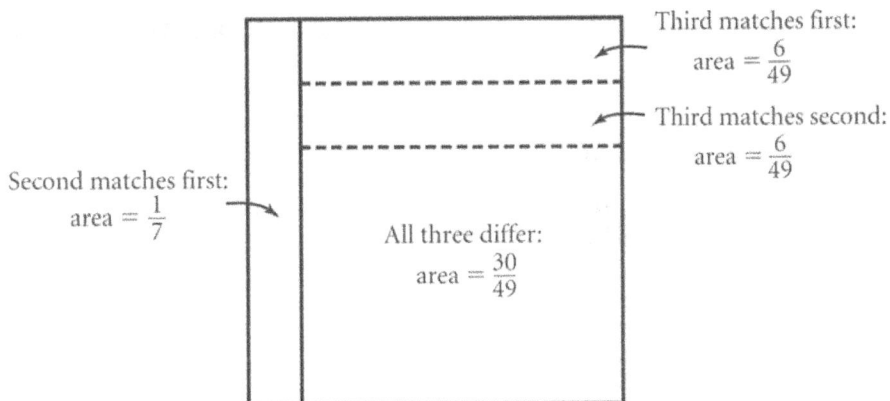

For four people, the big section (with area $\frac{30}{49}$) must now be subdivided to indicate whether the fourth person matches any of the first three. Students should explain that in the resulting diagram, the newly created sections for the three possible matches are each $\frac{1}{7}$ of the previous $\frac{30}{49}$, so each of these new sections has an area of $\frac{30}{343}$. The area of the "All four differ" section is the remaining $\frac{4}{7}$ of this $\frac{30}{49}$, so its area is $\frac{120}{343}$.

The final area diagram might look like this:

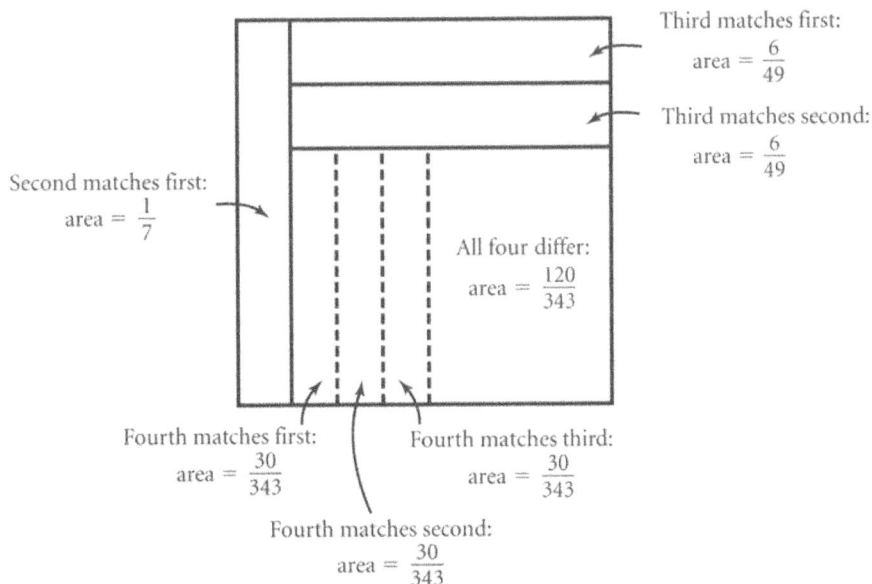

You may want to bring out that in this diagram, "Fourth matches first" is short for "Fourth matches first, and first three all differ," and so on. You also may want to have students verify that the areas of the individual sections of the diagram add up to 1.

Explaining Without a Diagram

Students who are comfortable with the principle of multiplying probabilities may be able to explain the problem without reference to any type of diagram.

Ask, **How can you find the probability of a match without using a diagram?** You might suggest that they first find the probability of *not* getting a match.

For instance, they might reason that there are 7^4 ways altogether to assign days of the week to four people. If the four people are to be born on different days, there are 7 choices for the first person, then 6 remaining choices for the second person, 5 remaining choices for the third person, and 4 remaining choices for the fourth person. This gives a total of $7 \cdot 6 \cdot 5 \cdot 4$ possible ways to assign distinct birth days-of-the-week out of the total of 7^4 ways. Because all ways are assumed to be equally likely, the probability of all four people being born on different days is $\frac{7 \cdot 6 \cdot 5 \cdot 4}{7^4}$, which simplifies to $\frac{120}{343}$.

Another way to express this approach is to see that there are 6 ways out of 7 for the first two to differ, and then 5 ways out of 7 for the third to differ from the first two, which makes $6 \cdot 5$ out of 49 ways for the first three to differ. Then there are 4 ways out of 7 for the fourth person to differ from the first three, which makes $6 \cdot 5 \cdot 4$ ways for all four to differ out of 343 ways to assign days of the week to the second, third, and fourth people. That gives a probability of $\frac{6 \cdot 5 \cdot 4}{7^3}$. Of course, this fraction is merely a simplified ("reduced") form of the fraction in the previous paragraph.

You can ask students to relate this approach to the bottom branch of the tree diagram, which shows that the probability of "no match" is simply the product $\frac{6}{7} \cdot \frac{5}{7} \cdot \frac{4}{7}$. The principle behind multiplying the probabilities is essentially the same as the principle for multiplying the numbers of cases.

Confirming Day-of-the-Week Matches

Tell students they will now get a chance to test out the theoretical probabilities for birth day-of-the-week matches. Review that for two people, the probability of a match is $\frac{1}{7}$, which is approximately .14. Inform students that they will pair up at random and see how many matches they get.

Before they do this, have them figure out how many pairs there will be. (You can include yourself if there is an odd number of students.) Then ask, How many of your pairs should have a match? Do you think you will actually get this number? How far off would the actual result have to be for you to doubt the theory?

Now have students pair off. You might randomize the pairing by having them count off by the number of pairs there are. Then have students with the same numbers get together and exchange birth days-of-the-week. Tally how many matches there are, and compare the result to the theoretical prediction.

Use this experiment to review the fact that in any random process, we expect some variation among the results. But also note that we can often predict how much variation is likely to take place.

Key Questions

Why does this multiplication process work?

How can you find the probability of a match without using a diagram?

How many of your pairs should have a match?

Do you think you will actually get this number? How far off would the actual result have to be for you to doubt the theory?

Monthly Matches

Intent

In this activity, students continue to work with conditional probability.

Mathematics

This activity asks students to consider probabilities that a number of people chosen randomly will include at least two individuals who were born in the same month. Once again they apply the principle that the probability of an event *not happening* is 1 minus the probability of the event *happening*.

Progression

Monthly Matches gives students one more experience with the ideas discussed in *Day-of-the-Week Matches* and *"Day-of-the-Week Matches" Continued* before attacking the "real" birthday problem.

Approximate Time

30 minutes for activity (at home or in class)

10 minutes for discussion

Classroom Organization

Individuals, followed by whole-class discussion

Doing the Activity

Students should be able to complete this activity independently.

Discussing and Debriefing the Activity

Question 1 provides an opportunity to review the pigeonhole principle. Students should see that they need 13 people to be certain of a match. (As needed, review that "100 percent probability" means being certain of the result.)

Use Question 2 to emphasize the assumption that each month should be considered equally likely. Thus, the probability for a match between two people is $\frac{1}{12}$.

You might want to have more than one presentation on some of the subsequent cases. Students should see that the reasoning is the same as for day-of-the-week matches. They will probably also see that it's easier to figure out the probability of not matching and then subtract from 1 to find the probability of matching. Here are the probabilities:

- For three people, the probability of not having a match is $\frac{11}{12} \cdot \frac{10}{12}$, which is about .76, so the probability of at least one month-of-birth match is about .24.

- For four people, the probability of not having a match is $\frac{11}{12} \cdot \frac{10}{12} \cdot \frac{9}{12}$, which is about .57, so the probability of at least one month-of-birth match is about .43.

- For five people, the probability of not having a match is $\frac{11}{12} \cdot \frac{10}{12} \cdot \frac{9}{12} \cdot \frac{8}{12}$, which is about .38, so the probability of at least one month-of-birth match is about .62.

Thus, there must be a minimum of five people for the probability of a month-of-birth match to be at least $\frac{1}{2}$.

The Real Birthday Problem

Intent

In this activity, students apply their recent work to a classic mathematics problem.

Mathematics

The Real Birthday Problem asks students to guess and then calculate the minimum number of people needed so that the probability of having at least two of them with the same birthday is greater than $\frac{1}{2}$.

Progression

This assignment culminates the thread that began with *Day-of-the-Week Matches.*

Approximate Time

25 to 30 minutes for activity

10 to 15 minutes for discussion

Classroom Organization

Pairs, followed by whole-class discussion

Doing the Activity

You may want to state that the question in the next activity is a classic mathematics problem, with an answer that is quite surprising to most people. It may be interesting to have students read the problem and share their guesses with the whole class before they actually start working on Question 2.

Have students work in pairs on this assignment, with one partner doing the calculator work and the other doing the recording. When both pairs in a group are done, they can share ideas.

If students seem stuck, you might suggest that they start with a specific case, such as finding the probability that there is a birthday match in a group of seven people. If needed, remind them of the technique of first finding the probability of not getting a match.

If any groups finish early, have them calculate the probability that at least two members of *their class* have the same birthday. (More precisely, they should find the probability of a match in a randomly chosen group of that size.)

Note: Students might run into calculator problems with this assignment. For instance, suppose they represent the probability of no birthday match among four people as the product $\frac{364}{365} \cdot \frac{363}{365} \cdot \frac{362}{365}$. If they continue in this way and try to compute the numerator and denominator of the product separately, the values may become too large for the calculator to handle. One solution is to calculate each result from the previous one. For example, to get the probability of no match for five people, multiply the probability for four people by 361 and then divide that result by 365.

Discussing and Debriefing the Activity

Have one or more students present the solution to Question 2. It may help students if they compile their results into a table, using the technique of finding the probability of a match by first finding the probability of not having a match. Such a table might look like this, which shows that a minimum of 23 people are needed for the probability of a match to be at least $\frac{1}{2}$:

Number of people	Probability of no match	Probability of at least one match
2	$\frac{364}{365}$ ($\approx$.997)	$\approx$ .003
3	$\frac{364}{365} \cdot \frac{363}{365}$ ($\approx$.992)	$\approx$ .008
4	$\frac{364}{365} \cdot \frac{363}{365} \cdot \frac{362}{365}$ ($\approx$.984)	$\approx$ .016
. . .	. . .	. . .
22	$\frac{364}{365} \cdot \frac{363}{365} \cdot \frac{362}{365} \cdot K \cdot \frac{344}{365}$ ($\approx$.524)	$\approx$ .476
23	$\frac{364}{365} \cdot \frac{363}{365} \cdot \frac{362}{365} \cdot K \cdot \frac{343}{365}$ ($\approx$.493)	$\approx$ .507

Probability for the Class

Once students have found the minimum number of people for the probability of a match to reach $\frac{1}{2}$, ask, **What is the probability that there would be at least one birthday match in a group the size of this class?** (If some students have already done this, let them present their result.) For instance, for the case of 30 people, the table would give a probability of .706 that there will be at least one birthday match.

Ask, **What does this probability mean?** For instance, someone might explain that if you looked at ten classes of 30 students each, about seven of those classes would have a match and three would not.

Students will probably be curious as to whether there is a birthday match among themselves. After determining whether there is, discuss what this means in terms of the theoretical analysis that the class has completed. Be sure students realize that the result for their particular class can neither confirm nor refute the general theory.

To reinforce the concept, you might have the class identify a group of classes at your school with roughly the same number of students in each and see whether they fit the theoretical probabilities. For instance, your class might survey 10 classes with about 30 students in each to find out how many of them include a birthday match. Then discuss the results, noting that in any simulation like this, the actual results may not match the expected value.

Comparing the Answer to Students' Guesses

Now that the students know the answer to the birthday problem, ask, **How does the answer compare to your original guess?** Very likely, many of them thought that the answer should be about half of 365, that is, about 183. (For days of the week, four people were needed, and for months of the year, five people were needed. Both of these results are close to half the number of possible outcomes.)

Students may be amazed that the actual answer, 23, is so much smaller. You can tell them that most people are surprised when they first learn the answer to the birthday problem.

Teacher Background: Why is the Answer to the Birthday Problem Surprising?

When people guess "about 183," they are likely thinking about the following similar-sounding problem, which we might call the *personalized birthday problem*:

What is the minimum number of people you need so that the probability of having at least one of them match your birthday is greater than $\frac{1}{2}$ *?*

Actually, the answer to this question is not 183 either. With 183 people, the probability of a match is only about .395. But 183 is much closer to the right answer for the personalized birthday problem than it is for the "real" birthday problem.

To solve the personalized birthday problem, students need to find out when the probability of *not* getting a match is less than $\frac{1}{2}$. Because the probability of any given person not matching you is $\frac{364}{365}$, the probability of N people (chosen at random) all not matching you is $\left(\frac{364}{365}\right)^N$. Thus, students need to find out what the smallest exponent N is that will make $\left(\frac{364}{365}\right)^N$ less than $\frac{1}{2}$. The smallest such N is not 183, but 253.

More Teacher Background: What Question does "183" Answer?

There is a "birthday problem" for which 183 is the answer. The number 183 is how many people you need in the room for the *expected value* of the number matching your birthday to be more than $\frac{1}{2}$.

As pointed out previously, the probability that at least one person in a group of 183 people will match your birthday is about .395, but this probability includes cases in which there is more than one match. The proba bility of *exactly one* match is about .304, the probability of *exactly two* matches is about .076, and the probability of *exactly three* matches is about .013, and so on. This gives an expected value of approximately

$$(.304 \cdot 1) + (.076 \cdot 2) + (.013 \cdot 3) + \ldots$$

which sums to just over .5.

In other words, if you tested many groups of size 183, found the number of matches to your birthday in each group, and found the average number of matches, you should get a result slightly over .5. If you used groups with fewer than 183 people, the average would be under .5.

Key Questions

What is the probability that there would be at least one birthday match in a group the size of this class?

What does this probability mean?

How does the answer compare to your original guess?

Supplemental Activity

The Chances of Doubles (reinforcement) gives students another opportunity to apply the technique of finding the probability of an event happening by first finding the probability of the event *not* happening.

Six for the Defense

Intent

In this activity, students use tree diagrams to count the number of ways to get a given outcome.

Mathematics

Six for the Defense describes a situation in which students are chosen randomly for assignments in a mock trial on each of four days. Students are asked to calculate the probability of one student being chosen for various combinations of assignments.

Progression

This activity begins a new phase of the unit, in which students need to find the number of paths on a tree diagram that lead to a particular type of outcome.

Approximate Time

25 minutes for activity (at home or in class)

10 to 15 minutes for discussion

Classroom Organization

Individuals, followed by whole-class discussion

Doing the Activity

Students should be able to complete this activity independently.

Discussing and Debriefing the Activity

Begin by having a student present his or her tree diagram, and then ask for other ideas about the diagram. Point out that the diagram need not treat each possible outcome for the die as a separate branch. That is, students can combine dice rolls of 1 through 5 and label the resulting branch "Other role," because Mari only cares about how often she is a defense attorney. The diagram might look like this, where each stage represents one of the four days. (For simplicity, probabilities are only listed on individual branches for the first stage.)

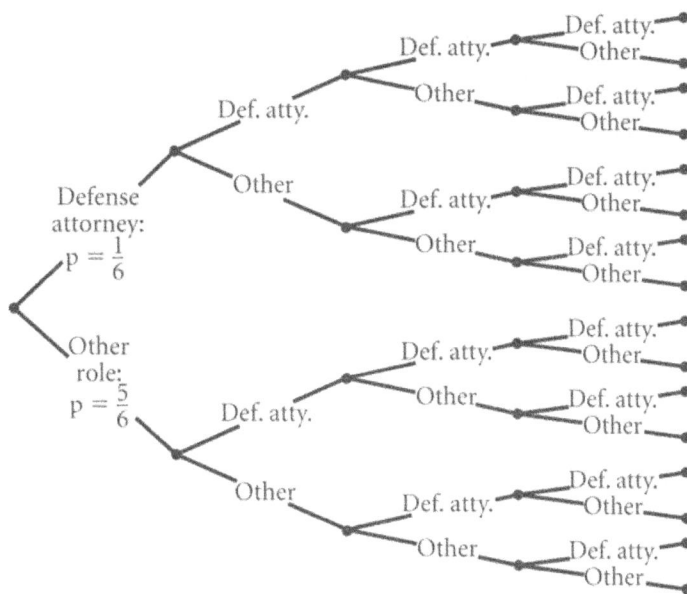

Questions 2 through 4

For each of Questions 2 through 4, there is only one path that leads to the desired outcome. In Questions 2 and 3, the subbranches along that path all have the same probability. Thus, Question 2 is represented by the top path in each case, which yields a probability of $\left(\frac{1}{6}\right)^4$, or approximately .0008. Question 3 is represented by the bottom path in each case, which yields a probability of $\left(\frac{5}{6}\right)^4$, or approximately .4823.

In Question 4, there is still only one path ("Other, Other, Def. atty., Other"), but the probabilities are not the same along that path. By multiplying the appropriate probabilities, students should see that Mari's probability of getting what she wants is $\frac{125}{1296}$, which is approximately .0965.

Question 5

Question 5 is the key element in this activity. Some students may think this is "obviously" four times the answer to Question 4, but for others it will not be so clear. The question requires students to consider that there are four possibilities for the day when Mari is defense attorney. In this case, that's fairly straightforward, but in other, similar-sounding counting problems, it will not be so easy. For now, have a student go through the tree diagram, path by path, and identify all of the paths involving only one "Defense attorney" day.

To find the probability that Mari is a defense attorney exactly once, students need to find the total of the probabilities for these paths. It's important that students recognize that these four paths each have the same probability, because they each involve three factors of $\frac{5}{6}$ and one factor of $\frac{1}{6}$. Students might add the four probabilities, or they might simply multiply $\frac{125}{1296}$ by 4.

Baseball and Counting

Intent

In this section, students learn to work with permutations and combinations.

Mathematics

As students work with calculating probabilities for the baseball situation from the central unit problem, they discover that most of the outcomes can occur in a number of different ways—the same number of games can be won and lost in many different orders. Being able to count the sequences that yield the same outcome is an important part of evaluating the probabilities related to the problem.

This section develops the concepts of combinations and permutations. Students see how the formulas for these values originate, how to calculate the number of combinations or permutations, and how to use the calculator to find those values without manual calculations. They learn how to distinguish between situations that involve combinations and those that involve permutations, and how the two values are related. Finally, they apply these principles to calculating probabilities for outcomes that can occur in a variety of ways.

Progression

Students begin this section by manually listing and counting the number of sequences that yield the same outcome using some of the simpler outcomes in the baseball problem, in *And If You Don't Win 'em All?* and *But Don't Lose 'em All, Either*. In these two activities, students find the probability that the Good Guys or Bad Guys win or lose all but one of their remaining games. In *The Good and the Bad*, they combine these probabilities to find probabilities for pairs of individual team outcomes occurring simultaneously, such as the probability that both the Good Guys and the Bad Guys win all but one of their remaining games.

Top That Pizza! moves to another analogy as students continue to list and count possible combinations, this time noticing the symmetry and relationships within their list.

The difference between combinations and permutations is represented vividly throughout this unit as the distinction between choosing scoops of different ice cream flavors to place in a bowl, where the sequence does not matter, and choosing flavored scoops to put onto an ice cream cone, where someone might prefer they be stacked in a particular order. *Double Scoops* begins with the distinction between two-scoop bowls and two-scoop cones, and in *Triple Scoops*, students develop a procedure for finding the number of permutations for larger cones. *More Cones for Johanna* requires students to generalize further by varying the number of flavors from which to choose.

Because the formula for the number of permutations (cones) is much easier to develop than that for the number of combinations (bowls), an important facet of this unit is recognizing the relationship between the number of permutations and the number of combinations. Students explore this relationship in *Cones from Bowls, Bowls from Cones* and *Bowls for Jonathan*.

Students apply these principles in yet another context in *At the Olympics* and are then introduced to the vocabulary and notation of permutations and combinations. They convert the procedures and relationships they developed earlier into formulas for calculating these values in *Formulas for $_nP_r$ and $_nC_r$*.

After practice with applying their new formulas (*Who's on First?*) and differentiating between combinations and permutations (*Which is Which?*), students return to the central unit problem with *Five for Seven* and *More Five for Sevens*, and then are shown how to use the graphing calculator to evaluate $_nP_r$ and $_nC_r$.

And If You Don't Win 'em All?

But Don't Lose 'em All, Either

The Good and the Bad

Top That Pizza!

Double Scoops

Triple Scoops

More Cones for Johanna

Cones from Bowls, Bowls from Cones

Bowls for Jonathan

At the Olympics

POW 12: Fair Spoons

Which is Which?

Formulas for $_nP_r$ and $_nC_r$

Who's on First?

Five for Seven

More Five for Sevens

And If You Don't Win 'em All?

Intent

In this activity, students find the probability that the Good Guys or the Bad Guys win all but one of their games.

Mathematics

The activities now begin to focus on situations where students must not only find the probability of a sequence of events occurring, but must also consider the number of equivalent sequences, such as where the same events occur in a different order.

Progression

The discussion focuses on compiling an orderly list of the possible sequences of wins and losses that yield the desired record.

Approximate Time

25 minutes for activity

10 to 15 minutes for discussion

Classroom Organization

Small groups or individuals, followed by whole-class discussion

Doing the Activity

This activity builds on the ideas from *Six for the Defense*. Before groups begin, you may want to review how to find the probability of each team winning all seven of its remaining games.

In Question 1, students may have trouble seeing that the probability is $.62^6 \cdot .38$. If so, you might suggest that they construct at least part of a tree diagram for the situation or write a list of the outcomes in order (that is, write W W W W W W L). These approaches may help students see that they need to multiply probabilities for seven games. For the first six games, the Good Guys are supposed to win, so the probability for each of these games is .62. Because they are supposed to lose the last game, the probability for that game is .38.

As groups work on Questions 2 and 3, help them to see that there are two aspects to these problems.

- Finding the probability of any particular sequence of results with a given number of wins and losses

- Finding the number of such sequences

As a hint on Question 2, you might ask how many different paths there are to a final record of six wins and one loss. You can suggest that groups make a tree diagram to illustrate the possible sequences or simply create a list of possible sequences of results.

Discussing and Debriefing the Activity

Begin by having a student present the solution to Question 1. The presenter may want to use a partial tree diagram or a diagram like this to explain that this sequence has a probability of $.62^6 \cdot .38$.

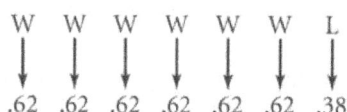

$$
\begin{array}{ccccccc}
W & W & W & W & W & W & L \\
\downarrow & \downarrow & \downarrow & \downarrow & \downarrow & \downarrow & \downarrow \\
.62 & .62 & .62 & .62 & .62 & .62 & .38
\end{array}
$$

For Question 2, ask, **How many different paths are there to a final record of six wins and one loss?** Have a student share a list of all the possible sequences of wins and losses for the Good Guys that will result in exactly six wins. Such a list might look like this:

W	W	W	W	W	W	L
W	W	W	W	W	L	W
W	W	W	W	L	W	W
W	W	W	L	W	W	W
W	W	L	W	W	W	W
W	L	W	W	W	W	W
L	W	W	W	W	W	W

Be sure students understand that each of these sequences has the same probability (namely, $.62^6 \cdot .38$). Thus, the overall probability of the Good Guys winning exactly six games is $7 \cdot .62^6 \cdot .38$, which is approximately .1511. (We will give these probabilities to the nearest ten-thousandth, but that is a rather arbitrary decision.)

Elicit the observation that although each of these sequences is less likely than winning all seven games, their combined probability is more likely than that of winning all seven games. (Students found in *How Likely Is All Wins?* that the Good Guys had a probability of about .0352 of winning all their remaining games.) Thus, the Good Guys are more likely to end up with six wins and one loss than with seven wins.

Question 3 is identical to Question 2 except for the change of team. The probabilities of winning and losing for the Bad Guys are different from those for the Good Guys, so the answer is $7 \cdot .6^6 \cdot .4$, which is about .1306.

How Many "All Wins" Cases?

To lay the groundwork for discussion of special cases later on, ask, *How many cases are there in which the Good Guys win all of their games?*

Students should see that "of course" there is only one such case. You might bring out that they could write the probability for that case as $1 \cdot .62^7 \cdot .38^0$. This process is parallel to the method of computing the probability for the six-win/one-loss case as $7 \cdot .62^6 \cdot .38$ (and students might put an exponent of 1 on .38).

Counting the Cases

If it hasn't yet come up, ask, *How could you easily find the number of six-win/one-loss sequences?* If a hint is needed, ask, *How can you be sure you have them all?* Working with a list will also help. If students organize the list as shown previously, they should see that the "L" can go in any of seven places, so there are seven possible sequences to consider.

An analogous problem that may be more familiar to students involves the probability of getting exactly six heads if you flip seven coins. Any one of the seven coins could be the one that comes up tails, just as any of the seven games could be the one the Good Guys lose. This coin question may seem simpler than the baseball problem because every possible set of outcomes for the coins is equally likely, with probability $\left(\dfrac{1}{2}\right)^7$.

Point out to students that to solve the central unit problem, they will have to solve other problems similar to those in *And If You Don't Win 'em All?* In fact, tell them that what they have done so far are the "easy" cases—the remaining problems will be more difficult. You might suggest that they will find it easier to do these problems if they organize such lists in a systematic way to be sure that they have not omitted any possibilities. (Eventually, though, they will learn ways to count without making lists.)

Key Questions

How many different paths are there to a final record of six wins and one loss?

How many cases are there in which the Good Guys win all of their games?

How could you easily find the number of six-win/one-loss sequences?

How can you be sure you have them all?

But Don't Lose 'em All, Either

Intent

In this activity, students continue to work with counting the number of ways to get a given outcome.

Mathematics

But Don't Lose 'em All, Either asks students to find the probabilities of each team losing exactly six of their remaining seven games. This activity is conceptually the same as *And If You Don't Win 'em All?,* but it involves the opposite results.

Progression

Students work on the activity individually and then discuss their results as a class.

Approximate Time

25 minutes for activity (at home or in class)

5 minutes for discussion

Classroom Organization

Individuals, followed by whole-class discussion if needed

Doing the Activity

Students should be able to complete this activity independently.

Discussing and Debriefing the Activity

Use your judgment as to how useful presentations of the activity would be. You may simply want to have volunteers give their answers so individuals can check their results.

The probability that the Good Guys will win their first game and lose the remaining six is about .0019. The overall probability that the Good Guys will win exactly one game is about .0131. The overall probability that the Bad Guys will win exactly one game is about .0172.

The Good and the Bad

Intent

In this activity, students find the probability of two separate events both occurring.

Mathematics

In *And If You Don't Win 'em All?*, students found the probability that the Good Guys would win exactly six of their remaining seven games and the probability that the Bad Guys would do the same. Now they combine those probabilities to find the probability that both of these things will happen.

Then students are asked to find the probabilities for as many other combinations in their overall chart as they can from the information that they have so far.

Progression

Students work on the activity in groups and then discuss their results as a class.

Approximate Time

25 to 30 minutes for activity

10 to 15 minutes for discussion

Classroom Organization

Groups, followed by whole-class discussion

Doing the Activity

Ask, **How do the results from** *But Don't Lose 'em All, Either* **relate to your chart?**, referring to the charts they made for *Possible Outcomes* showing all the combinations of records for the two teams. Have them look at those charts and review how they found the probabilities for two of the cells in the chart, in *How Likely Is All Wins?* and *Diagrams, Baseball, and Losing 'em All*.

Tell students that in this activity, they will use their recent results to fill in more of the probabilities within their charts. You may want to post the probabilities for each team that they have found so far (including the results from *How Likely Is All Wins?*). They currently know these facts:

For the Good Guys:

- The probability of all wins is $.62^7$, which is approximately $.0352$.

- The probability of all losses is $.38^7$, which is approximately .0011.

- The probability of six wins and one loss is $7 \cdot .62^6 \cdot .38$, which is approximately .1511.

- The probability of one win and six losses is $7 \cdot .62 \cdot .38^6$, which is approximately .0131.

For the Bad Guys:

- The probability of all wins is $.6^7$, which is approximately .0280.

- The probability of all losses is $.4^7$, which is approximately .0016.

- The probability of six wins and one loss is $7 \cdot .6^6 \cdot .4$, which is approximately .1306.

- The probability of one win and six losses is $7 \cdot .6 \cdot .4^6$, which is approximately .0172.

Students should recognize that they can enter this information along the top row and down the left-hand column of the chart. If students need a hint to fill in the individual cells, review the idea of using an area or tree diagram to find the probability of the Good Guys getting a particular result *and* the Bad Guys getting a particular result.

Discussing and Debriefing the Activity

You might have a student from each group specify one cell of the chart for which they found a probability and give that value. If there is disagreement over any of these values, have the student present an explanation so the class can resolve the question.

At the end of this discussion, the chart might look like this, with 16 of the 64 cells filled in:

Good Guys' record for the final seven games

		7-0 p = .0352	6-1 p = .1511	5-2 p = ?	4-3 p = ?	3-4 p = ?	2-5 p = ?	1-6 p = .0131	0-7 p = .0011
Bad	7-0 p = .0280	G .0010	G .0042	G	T	B	B	B .0004	B .00003
Guys'	6-1 p = .1306	G .0046	G .0197	G	G	T	B	B .0017	B .0001
record	5-2 p = ?	G	G	G	G	G	T	B	B
for	4-3 p = ?	G	G	G	G	G	G	T	B
the	3-4 p = ?	G	G	G	G	G	G	G	T
final	2-5 p = ?	G	G	G	G	G	G	G	G
seven	1-6 p = .0172	G .0006	G .0026	G	G	G	G	G .0002	G .00002
games	0-7 p = .0016	G .00006	G .0002	G	G	G	G	G .00002	G .000002

Note: All values are rounded to the nearest ten-thousandth except values that are less than .0001, which are given to one significant digit. Because the probabilities for each team (the values at the left and top) are also approximations, values within the table may not always be the same as the rounded-off product of the values at the left and top.

Key Question

How do the results from *But Don't Lose 'em All, Either* relate to your chart?

Top That Pizza!

Intent

In this activity, students count combinations and recognize the symmetry in certain counting problems.

Mathematics

Top That Pizza! asks students to count the number of possible combinations of two of five pizza toppings and then of three of five toppings. They are then prompted to notice that these answers are the same, and are asked to explain why this special relationship holds.

The discussion elicits the observation that choosing two of the five toppings is the same as omitting three toppings. Finally, students consider the special cases of one topping, all the toppings but one, no toppings, and all the toppings.

Progression

This activity is the first in a series of activities involving counting the number of ways to choose things. Students will likely do the problems in this activity primarily by brute force—that is, by making lists of the possibilities.

Approximate Time

25 to 30 minutes for activity (at home or in class)

15 minutes for discussion

Classroom Organization

Individuals, followed by whole-class discussion

Doing the Activity

Students should be able to complete this activity independently.

Discussing and Debriefing the Activity

You might have students come to agreement in their groups on the answers to the three questions and then provide at least one explanation of each answer.

Question 1

There are two main approaches to Question 1, shown here as the "list" method and the "other" method. The main goal of the next several activities is to get students to see the connection between them.

We expect that most students will use the "list" method, so we touch on the "other" method only briefly. If many students focus on the "other" method, you should adapt the discussion to build on what students present.

The "list" method: The "list" method consists of simply listing all the possible combinations, preferably in a systematic way. Here is one way to organize the list of the ten possible two-topping pizzas:

> pineapple and olives
> pineapple and mushrooms
> pineapple and onions
> pineapple and anchovies
> olives and mushrooms
> olives and onions
> olives and anchovies
> mushrooms and onions
> mushrooms and anchovies
> onions and anchovies

Ask, **What patterns do you see in this list?** Encourage the observation that the list contains four combinations that begin with pineapple, three that begin with sausage, two that begin with mushrooms, and one that begins with onions, so the total number is 4 + 3 + 2 + 1. (This idea may be helpful to students when they work on Question 1 of *Double Scoops*.)

The "other" method: The "other" method involves counting each combination twice and then dividing by 2. For each of the five ingredients, there are four combinations involving that ingredient. (The four combinations involving pineapple are all shown in the previous list.)

This makes a total of twenty combinations. However, because this method counts each combination twice (for example, it lists both "pineapple and sausage" and "sausage and pineapple"), the answer is half of 20, which is 10.

This method can be clarified by thinking of 20 as the number of different pizzas if you care which of the two toppings is on top. *Double Scoops* makes this distinction explicit.

Question 2

Groups should have found that they got the same answer to both Question 1 and Question 2. This may be surprising, because each row of the list for Question 2 has more items than do the rows for Question 1. Also, each topping appears more times in the list for Question 2 than in the list for Question 1. A list for Question 2 could be organized like this:

pineapple, olives, and mushrooms

pineapple, olives, and onions

pineapple, olives, and anchovies

pineapple, mushrooms, and onions

pineapple, mushrooms, and anchovies

pineapple, onions, and anchovies

olives, mushrooms, and onions

olives, mushrooms, and anchovies

olives, onions, and anchovies

mushrooms, onions, and anchovies

Question 3

If no one has a good explanation for Question 3 (that is, for why Questions 1 and 2 have the same answer), give one yourself, along these lines:

When you choose two toppings for a pizza, you can also think of that as choosing three toppings to be left off. So every choice of three is associated with a unique choice for two, and vice versa. If Johanna always took the toppings that Jonathan left out, they would run out of combinations at the same time.

Ask, **How can you pair up the lists to show this?** Students should see that they need to match the top row of the first list with the last row of the second list, the second row of the first list with the next-to-last row of the second list, and so on. (Students often call such pairs the "pizza" and the "unpizza." It is sometimes easier to calculate the number of one type than the other, but the numbers are the same.)

Some students will understand this idea now and others will not—that's okay. This idea will be explored further later in the unit.

One Topping, n – 1 Toppings

To extend the ideas here, ask, **How many different choices of pizza would Jonathan have if he cut back to only one topping per pizza?** Because there are five toppings, there would be five choices.

Then ask, **What if Johanna went up to four toppings per pizza? How many choices would she have?** Students should be able to use the symmetry principle discussed previously to see that she again has five choices.

No Toppings, All the Toppings

Note: The notation $_nC_r$ and $_nP_r$ for counting combinations and permutations will be introduced in the discussion following *At the Olympics*, along with alternative notations for combinatorial coefficients. We use the notation $_nC_r$ here to clarify the discussion, but don't share it with students at this time.

As preparation for dealing later with the interesting issue of the meaning of the combinatorial coefficients $_nC_0$ and $_nC_n$, pose these questions to the class:

- **What if Jonathan decided not to get any toppings on his pizza? How many different pizzas could he get?**
- **What if Johanna decided to splurge and get all five of the toppings she likes on her pizza? How many different pizzas could she get?**

Students may find these questions silly (and they are), but they are the type of situation that $_nC_0$ and $_nC_n$ describe. The answer to each of the questions is 1, and we will ultimately define both $_nC_0$ and $_nC_n$ to be equal to 1. For now, you can deal with the questions at an intuitive level.

Key Questions

What patterns do you see in this list?

How can you pair up the lists to show this?

How many different choices of pizza would Jonathan have if he cut back to only one topping per pizza?

What if Johanna went up to four toppings per pizza? How many choices would she have?

What if Jonathan decided not to get any toppings on his pizza? How many different pizzas could he get?

What if Johanna decided to splurge and get all five of the toppings she likes on her pizza? How many different pizzas could she get?

Double Scoops

Intent

In this activity, students look at the importance of order in certain counting problems.

Mathematics

Double Scoops describes a shop with 24 flavors of ice cream. Students are asked to find the number of two-scoop bowls that could be made if each scoop must be a different flavor, and then the number of two-scoop cones, if the order of the scoops upon the cone is important.

In the discussion, students note that making a list is not practical and that there are twice as many different two-scoop cones as there are two-scoop bowls.

This activity begins to lay the groundwork for differentiating between combinations and permutations, although that vocabulary is not yet introduced.

Progression

Students work on the activity in groups and then discuss their results as a class.

Approximate Time

25 minutes for activity

10 minutes for discussion

Classroom Organization

Groups, followed by whole-class discussion

Doing the Activity

No formal introduction of *Double Scoops* is necessary. Students may begin by trying to make lists to answer the two questions, but they will probably see that this is rather cumbersome and look for a shortcut, building on observations from *Top That Pizza!*

Using hints based on the analogous pizza problem, encourage groups to organize their lists, or at least the beginnings of a list, until the point at which they see a pattern. A good idea is to build on the $4 + 3 + 2 + 1$ pattern mentioned in the pizza problem discussion, to see that the answer here is $23 + 22 + \ldots + 1$.

Discussing and Debriefing the Activity

The distinction between *bowls* and *cones* is a key idea of the unit, and you should be sure students see this distinction.

Before discussing the two questions, ask, **Which answer should be larger?** Have a volunteer explain his or her answer. Apart from the numerical details, students should see that there are more *cones* than *bowls*. This fundamental intuition will be helpful throughout this unit.

On Question 1, elicit the observation that making a complete list is not a practical approach. As already noted, a good way for students to count the number of possible cones is to see the total as the sum $23 + 22 + \ldots + 1$, which equals 276.

On Question 2, students will probably realize that the answer must be exactly twice the answer to Question 1. That is, for each bowl that Jonathan can make, Johanna can make two cones by arranging the two flavors in either order. Make this relationship explicit in the discussion, as it's an important element in the overall development.

Note: Students may find the answer to Question 2 by some other method. For instance, they may see that there are 24 choices for the bottom flavor and then 23 choices in each case for the top flavor, giving $24 \cdot 23$ total choices. If they use an approach like this, note that the result, 552, is exactly twice the answer to Question 1. Then have someone explain why the two answers should have this relationship.

Key Question

Which answer should be larger?

Triple Scoops

Intent

This activity extends the ice cream situation from two scoops to three.

Mathematics

Triple Scoops asks students to find the number of three-scoop cones that could be made from the 24 flavors, and then the number of four-scoop cones. Students begin to recognize the relationship between the numbers of cones that are possible with successive numbers of scoops. This relationship is expressed in terms of a factorial.

Progression

The situation from *Double Scoops* is expanded to three- and four-scoop cones.

Approximate Time

20 to 25 minutes for activity (at home or in class)
15 to 20 minutes for discussion

Classroom Organization

Individuals or small groups, followed by whole-class discussion

Doing the Activity

Students should be able to complete this activity independently.

Discussing and Debriefing the Activity

The simplest approach to Question 1 is to make an explicit list of the possibilities:

pistachio, boysenberry, chocolate
pistachio, chocolate, boysenberry
boysenberry, pistachio, chocolate
boysenberry, chocolate, pistachio
chocolate, boysenberry, pistachio
chocolate, pistachio, boysenberry

In other words, Joshua needed to bring Johanna six cones to be sure she got what she wanted. (Some students may find a tree diagram to be an easier way to visualize why there are six possibilities. Encourage a variety of explanations.)

Question 2

On Question 2, students may find it cumbersome to make a complete list. But encourage them to at least start one so they see how it might be organized.

One approach is to start with those cones that have butter pecan on the bottom. Students should see that there are six of these, corresponding to the six possibilities from Question 1, because the list for Question 1 gives the possible orders for the other three flavors. They can then see that if there are six cones with butter pecan on the bottom, there should also be six cones with pistachio on the bottom, six cones with boysenberry on the bottom, and six cones with chocolate on the bottom.

Another approach is to take one of the cones listed for Question 1 and look at the four places to insert the butter pecan scoop. (For the first cone in the list, butter pecan could be inserted at the bottom, between pistachio and boysenberry, between boysenberry and chocolate, or at the top.) Thus, each of the six sequences listed previously leads to four possible cones.

Using Factorials to Count Cones

The approaches just described for Question 2 focus on the fact that the number of four-scoop cones made from four specific flavors is four times the number of three-scoop cones made from three specific flavors. The next stage in the analysis is for students to see that they can express the number of four-scoop cones directly using factorials.

You might accomplish this by asking, **How is the number of four-scoop cones related to the number of three-scoop cones?** That is, elicit an equation such as:

number of four-scoop cones = (4 · number of three-scoop cones)

Then ask for an analogous equation relating the number of three-scoop cones to the number of two-scoop cones, and then (somewhat trivially), an equation relating the number of two-scoop cones to the number of one-scoop cones.

Use these ideas to build a chain of equations something like this, concluding with the fact that there is only 1 one-scoop cone.

$$\text{number of four-scoop cones} = 4 \cdot (\text{number of three-scoop cones})$$
$$= 4 \cdot 3 \cdot (\text{number of two-scoop cones})$$
$$= 4 \cdot 3 \cdot 2 \cdot (\text{number of one-scoop cones})$$
$$= 4 \cdot 3 \cdot 2 \cdot 1$$

Use this discussion as an opportunity to review the concept of factorials and the notation n! as a shorthand for $n \cdot (n - 1) \cdot \ldots \cdot 2 \cdot 1$. Bring out that just as the number of four-scoop cones is 4!, so the number of three-scoop cones is 3!, and so on down the line.

Students should see that in general, they can make n! different cones from n scoops of different flavors.

Explaining Factorials with Trees

Ask, **How can you use a tree diagram to explain the connection between factorials and counting cones?** Help students see that with a tree diagram, they can picture the forming of a four-scoop cone as a four-step process. For instance, the diagram can be drawn showing four initial branches (for choosing the bottom flavor), three subbranches for each initial branch (for choosing the next flavor), and so on.

Key Questions

How is the number of four-scoop cones related to the number of three-scoop cones?

How can you use a tree diagram to explain the connection between factorials and counting cones?

More Cones for Johanna

Intent

In this activity, students formulate a rule for counting the number of possible permutations when several objects are selected from a pool of a given size in a situation where order is significant.

Mathematics

In the discussion following *Triple Scoops*, students found an expression to describe the number of different cones of a given size that could be formed once the flavors had been selected. *More Cones for Johanna* asks how many different three-scoop and four-scoop cones could be made from the 24 flavors and then asks students to express this as a rule in terms of the number of scoops on the cone. (Using vocabulary that has not yet been introduced, they will find a rule for the number of permutations of n objects taken r at a time. They will express this rule symbolically in *Formulas for $_nP_r$ and $_nC_r$*.)

Progression

Students work on the activity individually and then discuss their results as a class.

Approximate Time

25 to 30 minutes for activity

10 to 15 minutes for discussion

Classroom Organization

Individuals, followed by whole-class discussion

Doing the Activity

This activity continues the analysis of the number of ice cream cones. Students may need help seeing the distinction between the questions in this activity and the questions from *Triple Scoops*.

Note: In Questions 1 through 3, students need not get bogged down in translating "numbers of cones" into years, months, and days. Question 4 is included to provide a little twist for individuals that finish early.

Discussing and Debriefing the Activity

Have selected students offer their analyses for Questions 1 and 2, and encourage others to offer alternate explanations.

If students followed the tree diagram explanation for the use of factorials in the discussion of *Triple Scoops*, they will probably use a similar method here. They should see that Johanna has 24 choices for her bottom scoop, then 23 for the next scoop, and finally 22 for the top scoop. That's $24 \cdot 23 \cdot 22$ possible choices (12,144 choices altogether) for her three-scoop ice cream cones. That's a lot of ice cream—1 three-scoop cone each day for over 30 years!

The reasoning on Question 2 is the same as in Question 1, except for the additional factor of 21. Again, encourage students to offer varied explanations. Try to ensure that every student is comfortable with at least one way to explain that there are

$24 \cdot 23 \cdot 22 \cdot 21$ different four-scoop cones.

Question 3

Students may have a variety of ways of expressing a general rule in words or symbols. For instance, a student might say something like:

Start with 24 and keep multiplying by the number one less until the number of numbers being multiplied is equal to the number of scoops.

This is a good verbal description, although if the statement was phrased this way, you should elicit the word *factor* as a standard way to refer to "numbers being multiplied."

The symbolic representation of this idea will be developed in *Formulas for $_nP_r$ and $_nC_r$*, so you need not push now for such a formula. If students themselves push for such a representation, you can suggest that they use r to represent the number of scoops. The most difficult parts of the symbolic representation are using ellipses ("three-dot" notation) and recognizing that the last factor is $n - r + 1$ rather than $n - r$. (One way to avoid both problems is to write the expression as the quotient

$\dfrac{n!}{(n-r)!}$, but this shortcut should not be introduced now unless students suggest it.)

Question 4

As noted previously, Question 4 turns the situation around a bit. If any groups got to this question, you might have them present their solution. Students will probably have used guess-and-check to find the value of *n* for which $n(n - 1)$ is equal to 156.

Cones from Bowls, Bowls from Cones

Intent

In this activity, students see and apply the relationship between counting when order matters (permutations) and counting when order doesn't matter (combinations).

Mathematics

Cones from Bowls, Bowls from Cones asks students to determine the number of ice cream cones that can be made when given the number of bowls, and vice versa. Students will discover that the number of cones is $r!$ times the number of bowls (where r is the number of scoops).

Progression

This activity continues the theme from *Triple Scoops*.

Approximate Time

30 minutes for activity (at home or in class)

15 minutes for discussion

Classroom Organization

Individuals or small groups, followed by whole-class discussion

Doing the Activity

You may want to remind students at the start that in *More Cones for Johanna* they found a formula for the number of cones in terms of the number of scoops. To get a formula for the number of bowls in terms of the number of scoops, they must find a relationship between the number of bowls and the number of cones, because it's hard to get the number of bowls directly from the number of scoops.

Discussing and Debriefing the Activity

Have different students present each of Questions 1 through 4b, and then perhaps have a volunteer discuss Question 5.

On Question 1, students should see that there are two different two-scoop cones that correspond to the same two-scoop bowl, so there will be twice as many different cones as bowls. This means there are 930 different two-scoop cones at Francisco's Freeze.

A similar line of reasoning for Question 2 shows that the number of three-scoop cones at Paige's Parlor is six times the number of bowls. Thus, for three scoops, we have this relationship:

number of cones = 6 · (number of bowls)

Have the presenter review where the number 6 comes from. For instance, discuss why Jonathan's three-scoop bowl of ice cream led to exactly six different three-scoop cones for Johanna (in *Triple Scoops*).

Similarly, on Question 3a, review why 24 four-scoop cones can be made from each four-scoop bowl, especially the fact that the number 24 can be found as 4 · 3 · 2 · 1 (or 4!).

Use the discussion of Question 3b to establish the simple fact that the relationship from Question 3a can be reversed. Note explicitly that because the number of cones is 24 *times* the number of bowls, the number of bowls must be the number of cones *divided* by 24.

Because *Triple Scoops* did not involve five-scoop cones, you may want to spend extra time on Question 4a to ensure that students have generalized that the number of cones per bowl is the factorial of the number of scoops. Again, use Question 4b to establish the reversibility of the relationship between the number of cones and the number of bowls.

Question 5

The main goal in this problem is to establish the two-way relationship between the number of ice cream cones with a given number of scoops and the number of bowls of that size. Some students might express this directly in terms of factorials. For instance, if they use r to represent the number of scoops, they might write the answer to Question 5a as:

number of cones = $r!$ · (number of bowls)

Others may start with a more generic description of the relationship, such as

number of cones = (number of bowls) · (number of cones per bowl)

Some students may need the more generic description, so be sure to elicit that description, even if some students go directly to the factorial expression.

In Question 5b, bring out that one can simply reverse the relationship from Question 5a. This leads to the equation

$$\text{number of bowls} = \frac{\text{number of cones}}{\text{number of cones per bowl}}$$

You should post this principle. Be sure that students know how to find both the numerator and the denominator of this formula.

- For r scoops, the number of cones per bowl is $r!$.
- The number of cones of a given size is a "partial factorial" that starts with the number of flavors and has the same number of factors as the number of scoops under consideration.

You need not get a complete formula in terms of factorials here. That will come in *Formulas for $_nP_r$ and $_nC_r$*.

Bowls for Jonathan

Intent

In this activity, students apply the principles from *Cones from Bowls, Bowls from Cones.*

Mathematics

Bowls for Jonathan asks how many different three-scoop and four-scoop bowls can be made when choosing from 24 flavors of ice cream. Students begin to develop a formula for the number of ways of choosing *r* objects from a set of *n* objects.

Progression

This activity builds on the formula just developed in *Cones from Bowls, Bowls from Cones* and on *More Cones for Johanna*. The development of a formula for the number of ways of choosing *r* objects from a set of *n* objects will conclude in *Formulas for $_nP_r$ and $_nC_r$.*

Approximate Time

20 minutes for activity

15 minutes for discussion

Classroom Organization

Individuals, followed by whole-class discussion

Doing the Activity

Students should be able to complete this activity independently.

Discussing and Debriefing the Activity

The discussion of this activity should give you a good idea of how clear students are about the concepts from the past several activities. As with the discussion of *Double Scoops*, you might begin by asking whether there should be more bowls than cones or more cones than bowls.

Question 1

As students saw in *More Cones for Johanna*, there are $24 \cdot 23 \cdot 22$ different three-scoop cones (at the 24-flavor ice cream shop). Combining this with the fact that there are 6 three-scoop cones for each three-scoop bowl, the number of different three-scoop bowls is $\dfrac{24 \cdot 23 \cdot 22}{6}$, which equals 2024.

Keep in mind that many students will find it helpful to continue to state the relationship between bowls and cones in the multiplication form,

number of cones = 6 · (number of bowls)

as a separate step before going to the division form,

$$\text{number of bowls} = \frac{\text{number of cones}}{6}$$

Question 2

As with Question 1, answering this question is simply a matter of putting the components together. Students saw in *More Cones for Johanna* that there are 24 · 23 · 22 · 21 different four-scoop cones (at the 24-flavor ice cream shop), and they know that there are 24 four-scoop cones per four-scoop bowl. Therefore, there are $\frac{24 \cdot 23 \cdot 22 \cdot 21}{24}$ four-scoop bowls, for a total of 10,626.

At the Olympics

Intent

In this activity, students will apply the ideas from the past several activities to a new context.

Mathematics

In the context of selecting athletes for a team and assigning prizes to athletes, *At the Olympics* gives students a chance to apply their *bowls* and *cones* reasoning to situations that do not involve ice cream. The discussion following the activity introduces the terms **permutation** and **combination** and the related notation.

Progression

This activity involves several of the different counting principles that students have worked with so far. Students work individually, do a focused free-write in class, and then are introduced to new vocabulary and notation.

Approximate Time

20 minutes for activity (at home or in class)

10 minutes for focused free-writing

20 minutes for discussion

Classroom Organization

Individuals, followed by whole-class discussion

Doing the Activity

Students should be able to do this activity with no introduction. It is similar to the three-scoop ice cream problems, but with a different context.

Discussing and Debriefing the Activity

The main idea to elicit from this discussion is that Questions 1 and 2 are essentially like the ice cream problems for three scoops (with a total of ten flavors available).

You may want to collect this activity to get a sense of how well the class has understood the basic concepts of combination and permutation and their relationship to each other.

Focused Free-Writing: Cones versus Bowls

Today, students will be learning standard terminology and notation for concepts they've worked with for the past several days. In preparation for this, make a list of the relevant activities:

- *Top That Pizza!*
- *Double Scoops*
- *Triple Scoops*
- *More Cones for Johanna*
- *Cones from Bowls, Bowls from Cones*
- *Bowls for Jonathan*
- *At the Olympics*

Briefly review what each of these activities was about, and then have students do focused free-writing on this topic:

What do the questions in these activities have in common, and how do they differ from one another?

When students are done, give them an opportunity to share their thoughts. (Note: This focused free-writing will be a foundation for the upcoming activity *Which is Which?*)

Students should see that all of these problems involved counting how many ways to do things, and that they involved two different kinds of counting:

- One type in which they were simply picking a set of objects, including
    - choosing toppings for a pizza
    - choosing flavors for a bowl of ice cream
    - choosing a team to participate in the 400-meter dash

- A second type in which the objects were being selected in a particular order or in which each had a different role, including
    - choosing flavors for an ice cream cone
    - choosing winners for each of the gymnastics medals

Tell students they are about to learn standard terminology and notation for situations like these, and soon they will learn how to use graphing calculators to get answers to such problems.

Comment: Students have also looked at a third type of counting problem, in which one chooses one object from each of several sets, as in *Possible Outcomes*. Omit mention of this third type of problem now, as it further complicates an already complex idea.

Permutations

Begin the introduction of terminology and notation by reviewing how students counted the number of distinct three-scoop or four-scoop cones Johanna could make (in *More Cones for Johanna*). Then ask, **How many different nine-scoop ice cream cones can Johanna make?** (with 24 flavors available). Students should be able to explain that the answer is

$$24 \cdot 23 \cdot 22 \cdot 21 \cdot 20 \cdot 19 \cdot 18 \cdot 17 \cdot 16$$

Be sure students recognize that the product continues until there are exactly nine factors.

Tell students that problems of this type are called **permutation** problems and that this specific instance is called *the number of permutations of 24 objects taken 9 at a time.* Also inform them that this number is represented by the expression $_{24}P_9$ and that this expression is read as "twenty-four P nine."

Pose this question, asking students to express the answer in two ways—using the new notation, and as a product:

The pizza parlor serves five toppings that Johanna likes, and she wants to have three of these toppings on her pizza. Suppose she cares about the order in which the toppings go on. How many different pizzas can Johanna create?

They should see that the answer can be written both as $_5P_3$ and as the product $5 \cdot 4 \cdot 3$. In other words, $_5P_3$ represents the number 60.

Tell students that we often use n to represent the total number of objects (like the 24 flavors) and r to represent the number being selected and arranged (like the nine scoops). Thus, the general notation is $_nP_r$.

Combinations

Next, contrast Johanna's ice cream cones with Jonathan's bowls of ice cream. Tell students that problems like counting bowls of ice cream are called **combination** problems.

Give them a specific example, such as the number of three-scoop bowls of ice cream that can be made from 24 flavors, and tell them that this value is called *the number of combinations of 24 objects taken 3 at a time.*

Tell students that this number is represented by various notations and that they will be using two of them:

- $_{24}C_3$ (which is read as "twenty-four C three")

- $\binom{24}{3}$ (which is read as "twenty-four choose three")

Although $\binom{n}{r}$ may be the notation most often used traditionally, graphing calculators generally use $_nC_r$ (because it can more easily be written in a single line). These notations are used interchangeably in the IMP student materials.

Have a student describe how to figure out the numerical value of $\binom{24}{3}$. The student will probably go through a process similar to that used in *Bowls for Jonathan*. That is, there are $24 \cdot 23 \cdot 22$ different three-scoop cones, but there are six times as many three-scoop cones as three-scoop bowls, so there are $\dfrac{24 \cdot 23 \cdot 22}{6}$ different three-scoop bowls.

Pizza and Combinations

Review the pizza problem introduced earlier in the discussion to illustrate permutation notation. In that problem, students saw that if Johanna cared about the order of her toppings, the number of different three-topping pizzas could be written as $_5P_3$ or as $5 \cdot 4 \cdot 3$.

Then pose this similar question, again asking students to express the answer in two ways—using one of the new notations, and as a numerical value:

Jonathan can also choose from five toppings, but unlike Johanna, he doesn't care about the order of the toppings on his pizza. How many different three-topping pizzas can he create?

Students should see that they can express the answer as either $_5C_3$ or $\binom{5}{3}$. They should also explain why the numerical answer is 10.

Emphasize that the notations $_5C_3$ and $\begin{pmatrix} 5 \\ 3 \end{pmatrix}$ mean that we have five objects and want to know how many ways there are to choose three of the five. To make this concrete, you might list the ten possible three-topping pizzas again (as originally done for Question 2 of *Top That Pizza!*, where order didn't matter).

Tell students that these numbers are called **combinatorial coefficients** or **binomial coefficients**. As with permutations, the letters n and r are fairly standard.

More Examples

We suggest that you illustrate the notation with one or two more examples. For instance, ask how to represent how many different two-topping pizzas Jonathan could make if he didn't care about order (*answer:* $\begin{pmatrix} 5 \\ 2 \end{pmatrix}$ or $_5C_2$), or how many different six-scoop bowls of ice cream he could make at the 24-flavor shop (*answer:* $\begin{pmatrix} 24 \\ 6 \end{pmatrix}$ or $_{24}C_6$).

Permutations versus Combinations

Ask, **What is the difference between $_nP_r$ and $_nC_r$?** The key point to bring out is that with $_nP_r$, the order of the objects chosen is relevant, and with $_nC_r$, it is not.

Students might illustrate the distinction using examples. For instance, with Johanna's ice cream cones, she needs to know which flavor goes on the bottom, which next to the bottom, which on the top, and so on, so the number of cones is represented using $_nP_r$. Similarly, we need to know who gets each type of Olympic medal, so Question 2 of *At the Olympics* is a permutation problem.

On the other hand, the order in which scoops of ice cream are placed in the bowl doesn't matter to Jonathan, so the number of bowls is represented using $_nC_r$. Similarly, the three participants in Panacea's track team are not distinguished from one another, so Question 1 of *At the Olympics* is a combination problem.

Mnemonics

A variety of mnemonics have been invented for remembering which term refers to which concept, but we suggest that you let students make up their own. Be sure, however, that students know which word goes with which type of counting problem, because they will need that information in *Which is Which?*

By the Way

Here's a little mind-twister to think about in connection with permutations and combinations:

Isn't a "combination lock" misnamed? Shouldn't it be called a "permutation lock"?

Perhaps. For a lock in which the numbers go from 1 to 36 and one must dial three different numbers to undo the lock, the order of the numbers makes a difference, so there are $_{36}P_3$ different possibilities for the sequence. Thus, each possibility is really a permutation, not a combination.

Key Questions

How many different nine-scoop ice cream cones can Johanna make?

What is the difference between $_nP_r$ and $_nC_r$?

POW 12: Fair Spoons

Intent

In this activity, students explore a complex probability problem and then search for patterns in their results.

Mathematics

Fair Spoons returns to the situation from *Choosing for Chores*, in which Scott and Letitia determine who will wash the dishes by pulling two spoons from a bag. If the spoons are the same color, Scott washes, and if they are different colors, Letitia washes. Students are now asked to find combinations of numbers of spoons to place in the bag that will give each of the siblings an equal chance of having to wash the dishes. They are also asked to look for patterns in the combinations they find that yield equal probabilities.

Following the student presentations of the POW, the discussion introduces the phrase *sampling without replacement*.

Progression

Students work on the POW individually, and then several students present to the class.

Approximate Time

5 to 10 minutes for introduction

1 to 3 hours for activity (at home)

20 to 25 minutes for presentations and discussion

Classroom Organization

Individuals, followed by several student presentations and whole-class discussion

Doing the Activity

This POW builds on the problem introduced in Part I of *Choosing for Chores*. We suggest that you remind students of the result in that problem: If there are two purple spoons and three green spoons, there is a 40% chance of a match.

In this POW, they are looking for numbers of spoons of each color that will lead to a 50% chance of a match. Finding the general solution requires either seeing a pattern in the solutions or using an algebraic approach—the general solution forms a rather interesting pattern.

Suggestion: Many students jump to the conclusion that equal numbers of each color will lead to a 50% probability. You may want to go through a specific case as a whole class (such as five green spoons and five purple spoons) and post the result to eliminate this misconception.

Allow students about a week to work on this POW. On the day before the POW is due, choose three students to do POW presentations on the following day, and give them overhead transparencies and pens to take home to use for preparing the presentations.

Discussing and Debriefing the Activity

Have the selected students do their presentations. At the least, they should be able to present the probability of a match based on some numerical combinations other than the case of "two purple, three green" from *Choosing for Chores*.

It's hard to predict how far students will have gotten with the problem beyond such specific examples. They likely will have seen that making the numbers equal doesn't work. (They are especially likely to have realized this if you discussed one such example when the POW was assigned.)

Bring out that equal numbers of each color will never give a 50% chance of a match, because once the first spoon is selected, the remaining set of spoons no longer has the same number of each color. Emphasize that in the POW, the second spoon is chosen without replacing the first one, and tell students that this method of selection is called (naturally) *sampling without replacement*.

Elicit the observation that if the first spoon were put back in the bag before the second was chosen (which would be *sampling with replacement*), then starting with equal numbers of the two colors would always give a 50% chance of a match. (In fact, this is the only way to get a 50% chance of a match using sampling with replacement.)

Whatever results students got (for sampling without replacement), we suggest that you have the class compile a chart showing the probability of a match associated with each of the color combinations checked (including combinations that *do not* give a 50% probability). Such a table might help students find examples that give a 50% probability if they haven't yet done so or help them find a pattern among examples that they have already found.

Such a chart might look like this. (The result shown is from *Choosing for Chores*.)

Number of purple spoons	Number of green spoons	Probability of a match
2	3	.40

The simplest case that gives a probability of 50% is one purple spoon and three green spoons (or vice versa). In fact, there are infinitely many possibilities, and the solutions form an interesting pattern.

A Probability Formula for Spoons

One aspect of the problem that students might tackle as part of this discussion is to find a general formula for the probability of a match when there are P purple spoons and G green spoons. Even if no one did this, you can lead students through the reasoning based on the arithmetic of specific examples.

This tree diagram (from the discussion of *Choosing for Chores*) can be used for the general case by adjusting the numbers:

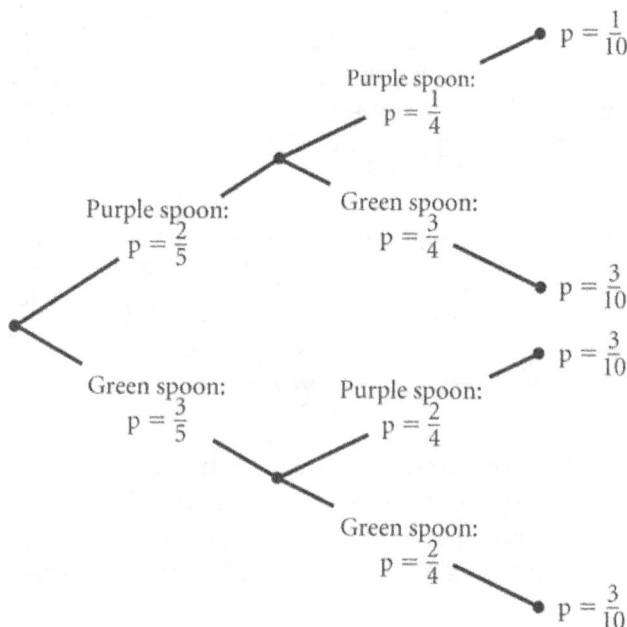

In the general case, the probability of choosing purple for the first spoon is $\dfrac{P}{P+G}$ instead of $\dfrac{2}{5}$, with similar adjustments for the other branches. Just as the probability of a match in the "2, 3 case" is $\left(\dfrac{2}{5} \cdot \dfrac{1}{4}\right) + \left(\dfrac{3}{5} \cdot \dfrac{2}{4}\right)$, so the general formula can be expressed in terms of P and G as

$$\left(\frac{P}{P+G}\right) \cdot \left(\frac{P-1}{P+G-1}\right) + \left(\frac{G}{P+G}\right) \cdot \left(\frac{G-1}{P+G-1}\right)$$

Setting this expression equal to $\frac{1}{2}$ gives an equation whose solutions solve the

problem. This equation can be simplified to $(G - P)^2 = G + P$. Among other things, this shows that the total number of spoons, $G + P$, must be a perfect square. (Notice that $P = 1$, $G = 3$ fits this equation.) Guessing and testing quickly yields a sequence of solutions (for bigger and bigger totals) which should look very familiar.

Which Is Which?

Intent

In this activity, students differentiate between combinations and permutations.

Mathematics

Which Is Which? asks students to review the activities they have done in this unit and to identify three problems that are combination problems and three that are permutation problems. They are then instructed to justify their selections and to express the answer to each using the proper notation.

Progression

In this activity, students apply new notation and terminology as they reflect on recent activities.

Approximate Time

30 minutes for activity (at home or in class)

10 minutes for discussion

Classroom Organization

Individuals, followed by brief group presentations

Doing the Activity

Students should be able to complete this activity independently.

Discussing and Debriefing the Activity

Have groups try to come to a consensus on whether each of the problems selected by their members is a permutation or a combination problem and on how to express the answer using the new notation. Have them make note of any disagreements.

Have a member of each group report any problems on which the group was unable to come to a consensus. If groups reached a consensus on all the problems, have them choose one to report on. Work with the class to get explanations that are as clear as possible. For instance, it is better for a student to say that an ice cream cone problem is a permutation problem "because a cone with chocolate on the top and vanilla on the bottom is different from a cone with vanilla on the top and chocolate on the bottom" than for the student to justify the category simply by saying, "because order matters."

Formulas for $_nP_r$ and $_nC_r$

Intent

In this activity, students develop formulas for $_nP_r$ and $_nC_r$.

Mathematics

This activity begins by having students look at specific cases expressed in $_nP_r$ notation and using these to develop a general formula for the number of permutations. Then, students consider the relationship between the number of permutations and the number of combinations and use that to develop a formula for the number of combinations in terms of n and r.

Progression

Three brief group presentations, one on Question 1, one on Question 2, and one on Questions 3 and 4, follow the activity.

Approximate Time

25 minutes for activity

15 minutes for discussion

Classroom Organization

Individuals, followed by several group presentations and whole-class discussion

Doing the Activity

This activity builds on the general relationship between $_nP_r$ and $_nC_r$ to get a formula for computing $_nC_r$. We suggest that you have the first group that finishes prepare a presentation on both Question 3 and Question 4, because the answer to Question 4 depends on the answer to Question 3. You can have other groups prepare presentations on Questions 1 and 2.

You need not wait until all groups have finished the activity before starting presentations, because the ideas behind the formulas have already been discussed in context. The key issue here relates to the abstraction of the concepts to symbols.

Discussing and Debriefing the Activity

On Questions 1a through 1c, ask students to refer to problem contexts with which to give meaning to the symbols. For example, $_nP_1$ can be interpreted as the number of ways to make a one-scoop ice cream cone when there are n flavors altogether, so $_nP_1 = n$.

Students looked at Question 2 in terms of ice cream cones in *More Cones for Johanna*. Here the problem is expressed more abstractly. They might state the answer like this:

$_nP_r$ *is the product* $n \cdot (n - 1) \cdot (n - 2) \cdot ...$, *continuing until you have r factors.*

At this stage of the unit, students should be able to express the last factor in terms of n and r. If needed, have them use a specific case such as Question 1c to see that the last term is n − r + 1 [or equivalently, n − (r − 1)] rather than n − r. In other words, students should develop the equation

$$_nP_r = n \cdot (n - 1) \cdot (n - 2) \cdot ... \cdot (n - r + 1)$$

Verifying the Formula using the Case r = n

You can use the special case in which r is equal to n as a way of confirming this formula. Ask, **What does this formula give as the value of $_nP_n$?** Students should see that it gives n! and should be able to justify this result in terms of the ice cream cone problem. If necessary, have students go back to Joshua in *Triple Scoops* and look at specific values of n. They should see that $_3P_3 = 6$ and that $_4P_4 = 24$. That should help them generalize to see that $_nP_n = n!$.

Question 3

For Question 3, have students explain how they used the specific situations to develop their equations. Most likely, they will use the idea that each r-scoop bowl leads to r! r-scoop cones as the basis for the equation

$$_nP_r = (r!) \cdot {_nC_r}$$

Note: Although some students may express $_nC_r$ in terms of $_nP_r$, probably most will do that only as part of Question 4.

Question 4

For Question 4, students should do two things.

Rewrite the equation from Question 3b to express $_nC_r$ in terms of $_nP_r$. That is, they should get the equation

$$_nC_r = \frac{_nP_r}{r!}$$

Substitute the expression they found in Question 2 for $_nP_r$.

Together, these two steps lead to the formula

$$_nC_r = \frac{n \cdot (n-1) \cdot (n-2) \cdot K \cdot (n-r+1)}{r!}$$

Students do not need to memorize this formula, especially because they will soon see how to find both $_nP_r$ and $_nC_r$ on the graphing calculator. (See *For Teachers: Computation versus Concepts* for a brief discussion of the pedagogical rationale for this emphasis.) Later in the unit, they will also see how to use Pascal's triangle as a technique for finding combinatorial coefficients (at least for small values of n). It is important, however, that they understand the connection between $_nC_r$ and $_nP_r$.

You may want to post either the final formula or the more conceptual expression giving $_nC_r$ in terms of $_nP_r$ (or both).

Be on the lookout for students who become overly dependent on these formulas and forget to think about what $_nC_r$ and $_nP_r$ mean. Periodic questions about the underlying ideas will help keep students focused on the concepts.

Key Question

What does this formula give as the value of $_nP_n$?

For Teachers: Computation Versus Concepts

There are two important elements in understanding combinations and permutations that should be distinguished from a pedagogical perspective.

- Knowing how to compute the number of combinations or the number of permutations
- Knowing which of the two concepts—combinations or permutations—to apply and knowing how to apply the appropriate concept

In this unit, we focus more attention on the second of these two elements. We caution against urging students to memorize formulas for computation. Rather, we suggest that they rely on posted reminders as well as the use of graphing calculators, especially for finding combinatorial coefficients.

On homework activities, for which students may not have graphing calculators available, students generally can simply express their answers in terms of the appropriate notation, rather than find actual numerical answers. In class, when graphing calculators are available, you should urge students to express answers both ways.

However, occasionally ask students to articulate the ideas behind the computations and to discuss the relationship between $_nP_r$ and $_nC_r$. In addition, periodically have them create a list, or at least the beginning of a list, to illustrate the set of objects being counted.

Who's on First?

Intent

This activity uses a baseball setting to illustrate counting principles.

Mathematics

The questions in this activity center around the number of different ways that players might be assigned to the various positions for a baseball team. The discussion focuses on the distinctions among the different counting principles.

Progression

Students work individually and then discuss their results as a class.

Approximate Time

25 minutes for activity (at home or in class)

10 minutes for discussion

Classroom Organization

Individuals, followed by whole-class discussion

Doing the Activity

Who's on First? uses several different counting principles in a single problem context.

Discussing and Debriefing the Activity

As volunteers present each problem, we recommend that you ask them to express their answers to Questions 1 through 3 both numerically and using either $_nP_r$ or $_nC_r$ notation. (In fact, none of the problems involves $_nC_r$, but you needn't mention that.)

For Question 1, students should get $7 \cdot 6 \cdot 5 \cdot 4$ (= 840) possibilities, which they should also express as $_7P_4$. (As the parenthetical comment at the end of Question 1 suggests, Sammy could more easily have put the names of the seven pitchers on slips of paper and drawn out four names, one at a time. But that's not how he selected his pitchers.)

For Question 2, the answer is $5 \cdot 4 \cdot 3$ (= 60), which can be written as $_5P_3$. You may want to have students explain why this is a permutation problem rather than a combination problem. Bring out, for instance, that an outfield of Willie Mays in left field, Joe DiMaggio in center field, and Mickey Mantle in right field is different from an outfield with the same players rearranged.

For Question 3, the answer is $6 \cdot 5 \cdot 4 \cdot 3$ (= 360), which can be written as $_6P_4$. If the distinction between $_nP_r$ and $_nC_r$ seemed clear to students in Question 2, you can probably skip discussion of this issue for Question 3.

Question 4 is a bit different, and it involves the multiplication principle used in *Possible Outcomes* (see the discussion of Question 3 from that activity). There are 7 choices for the first-game pitcher, 60 choices for the starting outfield, 360 choices for the starting infield, and only one choice for the catcher. So the number of possibilities is $7 \cdot 60 \cdot 360 \cdot 1$ (= 151,200).

Note: Some students may have used their answer to Question 1 in place of the number 7 in considering the pitcher choice. If so, this probably indicates that they didn't notice the reference to "the first game" in Question 4.

Finally, the answer to Question 5 is simply 9! (= 362,880), which students can also express as $_9P_9$.

Five for Seven

Intent

In this activity, students return briefly to the central unit problem, applying the ideas learned so far to a particular aspect of the problem.

Mathematics

Now that students know how to calculate combinatorial coefficients, they are able to consider one of the more difficult probabilities to calculate in the central unit problem, where the numbers of games won and lost yield more possible combinations than are easily visualized. In order to find the probability of a team winning a particular number of the remaining games, it is necessary to determine both the number of possible sequences that yield the correct number of wins and losses and also the probability of any one of those sequences occurring.

Progression

Students find the probability of the Good Guys winning exactly five of their seven games, and enter the new probabilities in their charts.

Approximate Time

25 minutes for activity

15 minutes for discussion

Classroom Organization

Groups, followed by whole-class discussion

Doing the Activity

As groups work on the activity, they should see that it has two aspects.

- Finding the probability of any particular sequence with the right number of wins and losses
- Finding the number of such sequences

(They dealt with the same pair of issues in *And If You Don't Win 'em All?*, but determining the number of sequences was simpler in that activity.)

Students will probably not have difficulty with the first issue. In Question 1, for instance, they should see that the probability of any particular sequence of five wins and two losses for the Good Guys is $.62^5 \cdot .38^2$. Therefore, the main focus of their work on this problem should be on determining the number of such sequences.

Although students are likely to use combinatorial coefficients to determine the number of sequences, we strongly recommend that you have them also make an explicit list of sequences (at least for Question 1) to confirm the correctness of their reasoning. (With the numbers in this activity, the list is of manageable size.) You may want to have a group prepare a transparency showing the 21 possible sequences.

Comment: Because the games are played one at a time, students sometimes think that the counting here involves permutations rather than combinations. You may need to help students to see, for instance, that "losing games 3 and 6" is the same outcome as "losing games 6 and 3." It may help if students imagine putting the numbers 1 through 7 in a bag and simply picking two slips to determine which two games the Good Guys lose.

Discussing and Debriefing the Activity

Begin the discussion of Question 1 by focusing on the counting aspect of the problem. This should be addressed both through an explicit listing of sequences and, more conceptually, through the use of combinatorial coefficients.

Counting with a List

Students might list all the sequences involving five wins and two losses in many ways. Remind students of the importance of organizing their lists in a way that will assure them that they have all the possibilities.

The list shown here is arranged by how soon the first loss occurs (starting with losing the first game) and, within each group, by how soon the second loss occurs.

L	L	W	W	W	W	W
L	W	L	W	W	W	W
L	W	W	L	W	W	W
L	W	W	W	L	W	W
L	W	W	W	W	L	W
L	W	W	W	W	W	L

W	L	L	W	W	W	W
W	L	W	L	W	W	W
W	L	W	W	L	W	W
W	L	W	W	W	L	W
W	L	W	W	W	W	L
W	W	L	L	W	W	W
W	W	L	W	L	W	W
W	W	L	W	W	L	W
W	W	L	W	W	W	L
W	W	W	L	L	W	W
W	W	W	L	W	L	W
W	W	W	L	W	W	L
W	W	W	W	L	L	W
W	W	W	W	L	W	L
W	W	W	W	W	L	L

In trying to understand this pattern, students might reason that when the first loss is in the first game, there are six possibilities for the second loss. When the first loss is in the second game, there are five possibilities for the second loss. When the first loss is in the third game, there are four possibilities for the second loss, and so on. So the number of possibilities is 6 + 5 + 4 + 3 + 2 + 1 = 21.

Counting with a Combinatorial Coefficient

Some groups may grasp the fact that the number of sequences is $_7C_5$. For others, however, this may need careful explanation, so be sure to discuss this idea thoroughly. As noted previously, some students may be unclear about why this is a combination problem and not a permutation problem.

Comment: This situation does not fit neatly into the "bowls and cones" metaphor that students have been using, but you may want to try to get them to connect the two situations. A useful analogy is to view the seven games as being like seven flavors. If the Good Guys are to lose two of their games, they need to select two of these seven "flavors." Because losing, for instance, games 3 and 6 is the same as losing games 6 and 3, this is a "bowls" problem rather than a "cones" problem. This may not be a totally satisfactory analogy, and students may find other explanations for why they want $_7C_5$ here and not $_7P_5$.

Finding the numerical value of $_7C_5$ may also not be a simple matter. (You should go over how to find this combinatorial coefficient even if students have already seen from an explicit list that the answer is 21.)

As needed, review the ideas behind the principle that $_nC_r$ can be found by dividing $_nP_r$ by $r!$. You may find it helpful to use the bowls-and-cones metaphor to explain this relationship. (After the previous discussion of *Who's on First?*, students will probably be clear about how to find $_nP_r$.)

Note: Students will see how to find $_nC_r$ and $_nP_r$ on a graphing calculator in the discussion following *More Five for Sevens*.

Completing Question 1

Once students are clear about why there are 21 different sequences by which the Good Guys win exactly five games, be sure they actually answer Question 1. That is, they need to combine the number of sequences with the fact that the probability for any one of these sequences of five wins and two losses is $.62^5 \cdot .38^2$. Thus, the probability that the Good Guys will win five and lose two of their remaining seven games is $21 \cdot .62^5 \cdot .38^2$, which is approximately .2778.

Question 2

Use the discussion of Question 2 to bring out that the number of sequences with two wins and five losses is the same as the number of sequences with five wins and two losses.

One way to see this is to take the original list of 21 sequences and replace the W's with L's and the L's with W's. You might also remind students of their work on *Top That Pizza!* In that problem, Jonathan was choosing two toppings out of five and Johanna was choosing three toppings out of five. Students saw that Jonathan and Johanna had the same number of choices.

Note: Students will see the principle that $_nC_r$ is equal to $_nC_{n-r}$ in connection with the upcoming activity *What's for Dinner?*, and again in connection with their work on Pascal's triangle. You may want to note this general principle now, though, in the context of this discussion.

Although the number of sequences is the same for Question 2 as for Question 1, the answers are different. Go over the two changes that students need to make:

- Replace the values .62 and .38 for the Good Guys with the values .6 and .4 for the Bad Guys.
- Interchange the exponents 2 and 5 from the earlier computation.

Thus, the probability that the Bad Guys will win exactly two of their remaining seven games is $21 \cdot .6^2 \cdot .4^5$, which is approximately .0774.

Question 3

If students have trouble finding the probability of both events happening together, you may need to remind them of their work in *The Good and the Bad*, in which they filled in many of the cells in their chart. As was done previously, you can have someone explain this using an area or tree diagram.

Students should see that they need to multiply the two probabilities, which gives a value of approximately .0215.

Entering the Values in the Chart

Have students get out their charts (last used in *The Good and the Bad*) of the probabilities for various outcomes for the two teams.

Discuss where they should enter their results from *Five for Seven*. Make sure they see that the answers to Questions 1 and 2 go along the top and the left sides, respectively, and that the answer to Question 3 goes in the appropriate cell within the chart.

More Five for Sevens

Intent

In this activity, students continue to find probabilities for combinations of multiple events as they fill in a few more cells on their charts for the central unit problem.

Mathematics

Student find the probabilities of several more cells in their chart. The discussion following this activity introduces finding combinatorial coefficients with the graphing calculator.

Progression

Five for Seven asked students to find the probabilities that the Good Guys win five games and lose two, that the Bad Guys win two of seven, and that both of those things happen. Now they are asked to find the probabilities for the other three situations that involve both teams winning or losing five games.

Approximate Time

25 minutes for activity (at home or in class)

20 minutes for discussion

Classroom Organization

Individuals, followed by whole-class discussion

Doing the Activity

Students should be able to do this activity independently, following the same steps as in *Five for Seven*.

Discussing and Debriefing the Activity

Although no new ideas are presented in this activity, it provides a good review of students' work in *Five for Seven*.

Have students check their answers with their group members, and then have students from different groups share their results with the class. As they do so, be sure students verify that they have the right values in their charts.

Their charts should now include all the values shown here:

Good Guys' record for the final seven games

		7-0 p = .0352	6-1 p = .1511	5-2 p = .2778	4-3 p = ?	3-4 p = ?	2-5 p = .0640	1-6 p = .0131	0-7 p = .0011
Bad	7-0 p = .0280	G .0010	G .0042	G	T	B	B	B .0004	B .00003
Guys'	6-1 p = .1306	G .0046	G .0197	G	G	T	B	B .0017	B .0001
record	5-2 p = .2613	G	G	G .0726	G	G	T .0167	B	B
for	4-3 p = ?	G	G	G	G	G	G	T	B
the	3-4 p = ?	G	G	G	G	G	G	G	T
final	2-5 p = .0774	G	G	G .0215	G	G	G .0050	G	G
seven	1-6 p = .0172	G .0006	G .0026	G	G	G	G	G .0002	G .00002
games	0-7 p = .0016	G .00006	G .0002	G	G	G	G	G .00002	G .000002

See *What About More Cells?* for a discussion of what to do if students want to solve the unit problem at this time.

Permutations and Combinations on the Graphing Calculator

Tell students they don't necessarily have to do the computations every time they want to find an $_nP_r$ or $_nC_r$ value. They can get the graphing calculator to do the arithmetic for them.

Give students a few minutes to explore how to do so. (You can let them consult their calculator manuals, give them hints, or simply tell them how to get these values.)

Have them confirm that the results they get from the graphing calculators agree with results they get by their own computations. For instance, they should verify that the calculators give 21 as the value of $_7C_2$ (used in *More Five for Sevens*) and 840 for the value of $_7P_4$ (used in Question 1 of *Who's on First?*).

Supplemental Activity

Determining Dunkalot's Druthers (reinforcement) involves reasoning similar to that needed in the unit problem, and you might suggest it at this time for students who are anxious to do more right now with these ideas.

What About More Cells?

Students may realize that they can fill in many more cells of the chart based on the information they now have, simply by multiplying probabilities they have along the top of the chart by probabilities along the left side. (For instance, they can find the probability for the cell that combines a record of seven wins and no losses for the Good Guys with a record of five wins and two losses for the Bad Guys.)

In fact, they may realize that they can now solve the unit problem, because the cases of three or four wins are not very different from the cases of five or two wins. However, the unit has some important topics to present before returning to solve the unit problem.

From a mathematics perspective, there's no reason students can't solve the unit problem now and then complete the rest of the unit afterwards. But students may remain more engaged in the unit if the central problem remains unsolved. Therefore, we suggest that you stick with the unit as organized unless students clamor to solve the problem now.

As a compromise, at this stage you might have them find only the probabilities for cells that use the results they already have. (That's an additional 16 cells beyond what's entered in the chart already.) That leaves 28 cells, involving three or four wins for each team, to be filled in for *Race for the Pennant! Revisited*.

Combinatorial Reasoning

Intent

In this section, students gain experience in applying combinatorial coefficients to a variety of probability problems.

Mathematics

As students were introduced to combinatorial coefficients through the activities in *Baseball and Counting*, the problems centered primarily around baseball (the theme of the central unit problem) and ice cream. While it is expected that the bowls versus cones analogy will continue of help students remember and make sense of the distinction between combinations and permutations, the activities in *Combinatorial Reasoning* utilize a greater variety of situations as students learn to apply these principles to probability problems.

In addition to application of combinatorial coefficients, several activities of this section also involve cumulative probabilities and evaluating a null hypothesis.

Progression

Nearly each activity in this section involves applying combinatorial coefficients in a new context, beginning with a counting problem in *What's for Dinner?* and moving on to probability problems in *The Perfect Group* and *Feasible Combinations*. Students use probability to evaluate a null hypothesis in *About Bias* and they consider cumulative probabilities in *Don't Stand for It* and *Stop! Don't Walk!*

All or Nothing examines special cases of $_nP_r$ and $_nC_r$ that are problematic in terms of the general formulas for those quantities. *Binomial Powers* prepares students to see the connection between combinatorial coefficients and binomial expansion in the next section, *Pascal's Triangle*.

What's for Dinner?

All or Nothing

The Perfect Group

POW 13: And a Fortune, Too!

Feasible Combinations

About Bias

Binomial Powers

Don't Stand for It

Stop! Don't Walk

What's for Dinner?

Intent

In this activity, students use permutation and combinatorial coefficients in context.

Mathematics

Students examine a new context involving both permutations and combinations. *What's for Dinner?* describes a situation in which a boy is tasked with providing a number of meals for his family each week, without repeating weekly menus. Because the directions he is given can be interpreted in two different ways (depending upon whether or not he is allowed to simply rearrange the order of the meals in a menu he has already used), the second question asks students to answer the first question again using the alternate interpretation.

The students discover in Question 3 that the value of $_7C_3$ is the same as that of $_7C_4$ from Question 1. The discussion following the activity brings out the general principle that $_nC_r = {_nC_{n-r}}$.

Progression

The questions in this activity lead students to focus on the difference between permutations and combinations.

Approximate Time

30 minutes for activity

25 to 30 minutes for discussion

Classroom Organization

Groups, followed by whole-class discussion

Doing the Activity

If groups need a hint for Question 1 of this activity, suggest that they compile a list of all the possibilities. (If they interpret the problem the way Lai Yee's parents do, there are only $\binom{7}{4} = 35$ combinations, so this is feasible.)

Question 3 is intended for groups that finish more quickly.

Discussing and Debriefing the Activity

The key element to look for in this discussion is whether students can distinguish clearly between $_nP_r$ and $_nC_r$.

Questions 1 and 2

Have students present Questions 1 and 2. They should see that under Lai Yee's interpretation, the answer to Question 1 is $_7P_4 \cdot \$20$, while under his parents' interpretation, the answer is $_7C_4 \cdot \$20$. (Remember that the question is "How much can he earn without repeating?," not simply "How many weeks can he go without repeating?")

Students might use graphing calculators to find that $_7P_4 = 840$ and $_7C_4 = 35$, but have them explain these numerical values in terms of the situation. For instance, a student might explain why $_7P_4$ is equal to 840 using a statement like "Lai Yee has seven choices for what to serve on Sunday. He then has six choices for Tuesday, and then five choices for Thursday, and then four choices for Saturday."

You might ask, **How could this problem be illustrated with a tree diagram?** Although they may object that 840 possibilities are too many to draw (and they are right), tell them that they merely have to start the tree and show how it would progress. For instance, they might use a schematic diagram like the following, explaining that at each stage the lower nodes would have branching just like that shown for the uppermost node at that stage.

Students can explain the value of 35 for $_7C_4$ by noting that from Lai Yee's point of view, there are 4! (= 24) different menus for each set of four meals, so $_7C_4$ should be equal to $_7P_4 \div 24$. (Students might also make a complete list of all 35 possible sets of four meals.)

Question 3

If students looked at their previous answers in terms of $_nP_r$ and $_nC_r$, they should have seen that adjusting from four nights to three simply means changing $_7P_4$ to $_7P_3$ and changing $_7C_4$ to $_7C_3$. Using Lai Yee's interpretation of the problem, this would substantially reduce his options for different menus, because $_7P_4 = 840$ and $_7P_3 = 210$. But there is no change in terms of his parents' interpretation, because $_7C_4$ and $_7C_3$ are both equal to 35.

Choosing Some is Excluding Others

Ask, **Why are $_7C_4$ and $_7C_3$ equal?** (This is the third context in which students have seen examples of the principle that $_nC_r = _nC_{n-r}$, so it may not surprise them. The principle came up previously both in *Top That Pizza!* and *Five for Seven*.)

Students might explain this specific example by pointing out that when Lai Yee chooses four meals, he is excluding three others, so the sets of four can be matched up with the sets of three. (This is essentially the same explanation that was used in the previous contexts.)

To move students closer toward the generalization that $_nC_r = _nC_{n-r}$, ask what combinatorial coefficient is numerically equal to $_{62}C_{17}$. Students should see that $_{62}C_{17}$ ought to be the same as $_{62}C_{45}$—the number 45 is found by subtracting 17 from 62.

Although students might articulate the general principle here that $_nC_r = _nC_{n-r}$, you need not push for it, because it will come up in *Combinations, Pascal's Way*.

Using the Formula to Explain $_7C_4 = _7C_3$

Students might gain further insight into why $_7C_4$ and $_7C_3$ are equal by looking at what the formula for $_nC_r$ gives in these two cases.

Ask, **How would you find each of the combinatorial coefficients $_7C_4$ and $_7C_3$ based on the formula?** They should come up with the expressions $\dfrac{7 \cdot 6 \cdot 5 \cdot 4}{4 \cdot 3 \cdot 2 \cdot 1}$ and $\dfrac{7 \cdot 6 \cdot 5}{3 \cdot 2 \cdot 1}$.

Then ask how they can know that the expressions are equal without finding either the numerators or the denominators separately. Students should see that they can "cancel" the 4's in the first fraction to get the second fraction.

Combinatorial Coefficients using Factorials

This is a good opportunity for students to see another way to write the formula for $_nC_r$. Point out that the numerator $7 \cdot 6 \cdot 5 \cdot 4$ (in the expression for $_7C_4$) is "part" of 7!, and ask how the fraction can be rewritten to show this. If necessary, suggest that students multiply both numerator and denominator by $3 \cdot 2 \cdot 1$. This gives the expression

$$\frac{7 \cdot 6 \cdot 5 \cdot 4 \cdot 3 \cdot 2 \cdot 1}{4 \cdot 3 \cdot 2 \cdot 1 \cdot 3 \cdot 2 \cdot 1}$$

Ask, **How can you write this new fraction using factorials?** They should see that it is simply

$$\frac{7!}{4! \cdot 3!}$$

Then ask how they could use this reasoning to write $_{62}C_{17}$ using factorials. It may be helpful to have students write $_{62}C_{17}$ out in full. You might start by asking what the last factor in the numerator should be, reviewing that it can be found as 62 – 17 + 1, which gives 46. Thus, $_{62}C_{17}$ is equal to

$$\frac{62 \cdot 61 \cdot 60 \cdot 59 \cdot 58 \cdot 57 \cdot 56 \cdot 55 \cdot 54 \cdot 53 \cdot 52 \cdot 51 \cdot 50 \cdot 49 \cdot 48 \cdot 47 \cdot 46}{17 \cdot 16 \cdot 15 \cdot 14 \cdot 13 \cdot 12 \cdot 11 \cdot 10 \cdot 9 \cdot 8 \cdot 7 \cdot 6 \cdot 5 \cdot 4 \cdot 3 \cdot 2 \cdot 1}$$

Help students to see that they can multiply the numerator and denominator by 45! to rewrite this more simply as

$$\frac{62!}{17! \cdot 45!}$$

Students should be able to see from this example that more generally, the combinatorial coefficient $_nC_r$ is equal to

$$\frac{n!}{r! \cdot (n-r)!}$$

Comment: When students see this formula, they may realize that it demonstrates that $_nC_r = {_nC_{n-r}}$, but as noted before, this generalization can wait until the discussion of *Combinations, Pascal's Way*.

Key Questions

How could this problem be illustrated with a tree diagram?

Why are $_7C_4$ and $_7C_3$ equal?

How would you find each of the combinatorial coefficients $_7C_4$ and $_7C_3$ based on the formula?

How can you write this new fraction using factorials?

All or Nothing

Intent

In this assignment, students explore the meaning of some special cases of $_nC_r$ and $_nP_r$.

Mathematics

Students consider the special cases $_nC_0$, $_nC_n$, $_nC_1$, $_nP_0$, and $_nP_1$. The discussion following the activity touches on why these cases require special consideration, and uses problem contexts to clarify the definitions.

Progression

Students revisit the context of problems from previous activities to help them make sense of some special cases of $_nC_r$ and $_nP_r$.

Approximate Time

30 minutes for activity (at home or in class)

25 minutes for discussion

Classroom Organization

Individuals, followed by whole-class discussion

Doing the Activity

If time allows, you may want to discuss why these cases require special treatment. For ideas, see the subsection "Why a Special Discussion," below.

Discussing and Debriefing the Activity

Let students discuss the first three problems in their groups for a few minutes. We recommend that you precede the discussion of this activity by talking about why it's necessary to deal with these cases in a special way, using ideas in the following subsection.

Why a Special Discussion?

Help students to see that the formulas for the coefficients are somewhat problematic in special cases. For instance, the expression

$$n \cdot (n - 1) \cdot (n - 2) \cdot \ldots \cdot (n - r + 1)$$

for $_nP_r$ usually has r factors, but that doesn't make sense if $r = 0$. We suggest that you bring out the analogy with the problem of defining 2^0. We can't use the usual definition of exponents, because that would mean "multiplying no factors of 2 together." Similarly, we need some special way to define $_nP_0$. Point out that we also need a special way for defining 0!, which is used to express the formal relationship between $_nC_0$ and $_nP_0$.

Acknowledge that defining these special cases of combinatorial and permutation coefficients in terms of situations is a bit problematic. For instance, there is something quite artificial in talking about "choosing 0 things out of n."

Tell the class that because of these complications, the special cases discussed in this activity need to be defined by some method other than the usual definition. We often describe such a special definition as a *convention*, although this does not mean that the definition is arbitrary. Students should see that the definitions used here (like the definition for 2^0) are those that are the most appropriate in terms of the various mathematical considerations.

Question 1

Have a student present Question 1. By analogy with the pennant race explanation of $\binom{7}{5}$, the combinatorial coefficient $\binom{7}{0}$ should represent the number of different sequences by which the Good Guys win no games. There is exactly one way for this to happen—the Good Guys need to lose seven games in a row—so $\binom{7}{0}$ is defined as 1.

Comment: In the discussion following *Top That Pizza!*, students looked at the question of creating a pizza with no toppings. You may want to remind them of that discussion and of their conclusion that there was exactly one no-topping pizza.

Question 2

Have another student discuss Question 2. The analogy here suggests that the combinatorial coefficient $\binom{5}{5}$ should give the number of pizzas Jonathan can create if he chooses all five of the toppings he likes. Because there is only one such pizza, we also define $\binom{5}{5}$ as 1.

Students may object to this definition on the basis that there really isn't *any* choice involved here, so the answer should be 0. You can acknowledge that there is some logic to this argument, but point out that when a person is given no choice, it generally means that there is only one thing that he or she can do. If you phrase the question as "the number of ways Jonathan can have a pizza with all five toppings" (rather than in terms of "choice"), students might feel more comfortable with this definition. You might also refer to the analogous $\binom{7}{7}$ for the baseball problem, asking, **How many sequences are there in which the Good Guys get exactly seven wins?** Elicit that there is exactly one sequence in which the Good Guys get exactly seven wins. Ultimately, though, you may need simply to tell students that this is the definition and they have to live with it (as they have probably already learned to live with the definition that $2^0 = 1$).

Comment: If students ask about the meaning of $\binom{5}{0}$ in the pizza context, explain that there is exactly one way to make a pizza with no toppings—simply leave all the toppings off it. (At many pizza stores, "no toppings" means a "plain cheese" pizza; the cheese itself is not considered a topping.)

Question 3

You may want to have two students each present one part of Question 3. Although $_{24}P_0$ and $\binom{24}{0}$ are both defined to be 1, be sure students distinguish between them. Johanna eats her ice cream on cones, where order matters, so for her, the relevant symbol is $_{24}P_0$. There is one way for her to get "no scoops," namely, to eat only the cone itself. Jonathan, however, eats his ice cream in bowls, where order doesn't matter, so for him, the relevant symbol is $\binom{24}{0}$. There is one way for him to get no scoops—an empty bowl.

Before leaving the ice cream problem, ask, **What's the numerical value of $_{24}P_{24}$?** (This was not part of the activity). Students should see that there are 24! ways for Johanna to arrange a cone of 24 scoops with 24 flavors This is approximately $6 \cdot 10^{23}$, which means that if she ate a million such cones every second, it would take about 20 billion years for her to try them all.

Summary of Questions 1 through 3

Use the examples discussed in Questions 1 through 3 to establish these general principles for every positive integer *n*.

- $\dbinom{n}{0} = 1$

- $\dbinom{n}{n} = 1$

- $_nP_0 = 1$

- $_nP_n = n!$

Emphasize that these principles are really definitions, because the situations on which $_nP_r$ and $\dbinom{n}{r}$ are usually based are somewhat artificial when r is either 0 or n, and the general formulas are problematic as well.

Also ask, **How do you think you should define** $_0P_0$ **and** $\dbinom{0}{0}$**?** It should seem reasonable that both expressions are defined to be 1, although neither expression makes much sense in the context of ice cream cones or baseball games. Point out that defining $_0P_0$ as 1 also seems to mean defining 0! as 1. (Make sure students realize that the usual definition of $n!$ doesn't work when $n = 0$.)

Finally, ask, **Are these definitions consistent with the equation** $\dbinom{n}{r} = \dfrac{_nP_r}{r!}$**?** The case that is potentially most problematic is $r = 0$. If we define 0! as 1 (as just discussed), then the relationship continues to hold true.

Question 4
Finally, call on volunteers to discuss Question 4. Students should be able to use one or more of the contexts to explain why, for any positive integer n, both $_nP_1$ and $\dbinom{n}{1}$ should be defined as equal to n.

You might ask, **Are there any other cases for which** $_nP_r$ **and** $\dbinom{n}{r}$ **are equal?**

(There aren't, and students should be able to articulate why not.)

Key Questions
How many sequences are there in which the Good Guys get exactly seven wins?

What's the numerical value of $_{24}P_{24}$**?**

How do you think you should define $_0P_0$ and $\begin{pmatrix} 0 \\ 0 \end{pmatrix}$?

Are these definitions consistent with the equation $\begin{pmatrix} n \\ r \end{pmatrix} = \dfrac{_nP_r}{r!}$?

Are there any other cases for which $_nP_r$ and $\begin{pmatrix} n \\ r \end{pmatrix}$ are equal?

The Perfect Group

Intent

This activity is an application of combinatorial coefficients.

Mathematics

In this activity, students use combinatorial coefficients to find a probability.

Progression

Students work on the activity individually and then discuss their results as a class.

Approximate Time

25 minutes for activity (at home or in class)

15 minutes for discussion

Classroom Organization

Individuals, followed by whole-class discussion

Doing the Activity

The Perfect Group asks students to find the probability of a student being randomly assigned to the exact four-person group that he would like, within a three-year span, if new groups are formed randomly every two weeks.

Discussing and Debriefing the Activity

You might begin by having people simply call out the guesses they made for Question 1 to see the variety of responses. (You should do this before the class has agreed on the "right" answer for Question 2.)

One aspect of Question 2 is figuring out how many different four-person groups there are that include Julio. Some students may think that this is $\binom{32}{4}$, which equals 35,960. But the problem only concerns groups that include Julio, so the appropriate number is $\binom{31}{3}$, which is 4495. (*Note:* See the next subsection, "Reviewing Combinatorial Coefficients," for ideas on how to get students to explain the value 4495.)

To come up with a probability for Question 2, students need to make some assumptions to determine how many groups Julio will be in over the course of three school years. They might figure about 40 weeks to a school year, which leads to about 20 groups a year and about 60 groups in the three-year period.

If Julio's groups were guaranteed to all be different, then his chances of getting his perfect group would be simply $\frac{60}{4495}$, which is about .013. Because the chances of his getting the same exact group twice are very small, this is a reasonable assumption.

If you don't want to make this assumption, the best way to analyze the problem is to say that Julio's probability of *not* getting his perfect group at any given time is $\frac{4494}{4495}$, and so the probability that he doesn't get his perfect group in 60 tries is. Therefore, the probability of getting his perfect group at least one of these 60 times is $1 - \left(\frac{4494}{4495}\right)^{60}$, which is also about .013. In fact, both answers, to four decimal places, give .0133, so not much is lost by the assumption.

Reviewing Combinatorial Coefficients

If it seems like a good time to have students review ideas about combinatorial coefficients, then ask, **How would you get the value of $\binom{31}{3}$ without using a graphing calculator or memorizing a formula?**

One approach is to assume that students are assigned to groups of four by playing cards; thus the kings of all four suits would form a group. Suppose that Julio had the king of clubs. There are 31 people who could have the king of spades, 30 remaining people who could have the king of hearts, and then 29 people who could have the king of diamonds. That gives $31 \cdot 30 \cdot 29$ possibilities. But then students need to take repetitions into account. If you take any particular combination—for instance, person A has the spade, B the heart, and C the diamond—then six different arrangements will give the same group of three. (You may want to have students actually list these six possibilities.) Thus, overall, there are $\frac{31 \cdot 30 \cdot 29}{3 \cdot 2 \cdot 1} = 4495$ possible groups that include Julio.

Key Question

How would you get the value of $\binom{31}{3}$ without using a graphing calculator or memorizing a formula?

Supplemental Activity

Sleeping In (reinforcement) involves ideas similar to those in *The Perfect Group*.

POW 13: And a Fortune, Too!

Intent

In this POW, students use algebra and logic to solve a complex problem.

Mathematics

This POW returns to the *Bags of Gold* situations from Year 1, although with a considerably more difficult problem.

Progression

If your students studied IMP Year 1 (particularly the unit *The Pit and the Pendulum*), this POW can be introduced with a brief discussion of how it differs from the similar POWs that they saw before. In any case, students need to be told to fully write up any partial solutions they obtain if they get stuck.

Give students about a week to work on the POW and then assign several students to make presentations. It is recommended that the solution be left open for further exploration if nobody manages to completely solve the problem.

Approximate Time

10 minutes for introduction

1 to 3 hours for activity (at home)

30 to 35 minutes for presentations and discussion

Classroom Organization

Individuals, followed by several presentations and whole-class discussion

Doing the Activity

This is a rather difficult problem, so be sure to encourage students to state any partial solutions they find and to indicate in their write-ups where they got stuck. (Students can always benefit from being reminded that it's better to know when you're stuck and be able to say so than to think you're not stuck when you are.)

On the day before the POW is due, choose three students to make POW presentations on the following day, and give them overhead transparencies and pens to take home to use for preparing those presentations.

Discussing and Debriefing the Activity

Ask students to make their presentations. Quite likely, no one came up with a complete solution, but ask if students had methods that would at least *sometimes* determine which bag had the counterfeit gold and whether it was heavier or lighter.

If not, ask what weighing schemes they tried. A fairly natural idea is to first weigh two of the bags and then weigh two others. It turns out that this is a successful way to start, although it may not be clear how to proceed after that.

If no one has a complete solution, we recommend that you leave this as an open problem. We provide a solution here but suggest that you use this information only as a source of possible hints, not as an outline for explaining the problem to the class.

A Possible Solution

Suppose we name the bags A, B, C, D, and E. We'll use the same letters to represent the weights of those bags.

We begin by weighing A and B together, and we call their combined weight x. Next, we weigh C and D and call their combined weight y. Symbolically, we have

$$x = A + B \text{ and } y = C + D$$

The analysis now breaks down into two cases.

Case 1: $x = y$. This is the same as saying that all four of these bags must be genuine, because otherwise the two pairs would have different weights. Thus, in this case, E is the bag with counterfeit gold. We can weigh it to find out if it is lighter or heavier than the genuine gold. (This is the easier of the two cases.)

Case 2: $x \neq y$. In this case, the counterfeit is one of the first four bags, and we know that E must contain genuine gold.

Here's where it gets subtle. A reasonable guess as to how to proceed would be to weigh E and thus find out the weight of a bag of genuine gold. Either x or y would be exactly twice that, and that would give us two more true bags. But we wouldn't know which of the other two bags was the counterfeit (although we would know if it was heavier or lighter than normal).

The trick is to do something more complicated than simply weigh E. Instead, we weigh a combination like B + D + E (combining E with one bag from each of the two previous pairs) and call this weight z. So we have $z = B + D + E$.

The counterfeit bag is either A, B, C, or D. We proceed by figuring out what would happen in each of the four cases. It turns out that in each case, there is a relationship involving x, y, and z that occurs only in that case. The cases in which A and C are counterfeit are similar to each other, and the cases in which B and D are counterfeit are similar to each other. We'll do A and C first.

Case 2a: A is counterfeit. In this case, bags B, C, D, and E all weigh the same, so $y = 2B$ and $z = 3B$, and we have $z = \frac{3}{2}y$. Thus, if A is counterfeit, then

$$z = \frac{3}{2}y.$$

Case 2c: C is counterfeit. This is just like case 2a. If C is counterfeit, then $z = \frac{3}{2}x$.

The cases of B and D are a bit more complicated.

Case 2b: B is counterfeit. Then A, C, D, and E are all the same, so the combined weight B + D + E is the same as $A + B + \frac{1}{2}(C+D)$. [You can think of this as saying B + D = A + B and $E = \frac{1}{2}(C+D)$.] In other words, if B is counterfeit, then $z = x + \frac{1}{2}y$.

Case 2d: D is counterfeit. Similarly, if D is counterfeit, then $z = \frac{1}{2}x + y$. Thus, in each of the four cases within case 2, we get one of the four equations $z = \frac{3}{2}y$, $z = \frac{3}{2}x$, $z = x + \frac{1}{2}y$, and $z = \frac{1}{2}x + y$.

The key to this analysis of the problem is that no more than one of these equations can hold. (If any two or more of them were true, then x and y would be equal. But in Case 2, x and y are different.) Thus, each of these equations is uniquely matched with one of the four cases within Case 2. By examining the numbers x, y, and z, we can determine which one of the four equations holds true and therefore which of A, B, C, and D is the counterfeit bag.

Once we know which bag is counterfeit, it is easy to tell if it is lighter or heavier by comparing x and y.

Supplemental Activity

Twelve Bags of Gold Revisited (extension) reexamines the Year 1 POW *Twelve Bags of Gold.*

Feasible Combinations

Intent

In this activity, students examine the use of combinatorial coefficients in the process of solving linear programming problems.

Mathematics

Feasible Combinations asks students to count the number of systems of equations that must be considered in order to solve linear programming problems with given numbers of variables, constraint equations, and constraint inequalities.

Progression

Although this activity contains no new mathematical ideas, it is included here to connect the concept of combinatorial coefficients to a context that has already acquired considerable meaning for students.

Approximate Time

25 minutes for activity (at home or in class)

10 minutes for discussion

Classroom Organization

Individuals, followed by whole-class discussion

Doing the Activity

Students should be able to do this activity with no introduction.

Discussing and Debriefing the Activity

Have volunteers present their solutions. You may want to discuss further why constraint equations should be included in the linear systems. Bring out that if one used a system that did not include these equations, its solution would have to be discarded anyway unless it happened to fit those equations.

Also point out that trying to solve all 924 systems (from Question 1a) would be an enormous waste of effort. It's much more efficient to check only the 28 systems that are needed in Question 1b (and even this number can be reduced by careful analysis).

Students might comment that a computer could do the work, so this would be a waste of effort. Point out that computer time costs money and that in real-world applications, the number of systems is often enormous, so every efficiency is worthwhile—and this is a major efficiency.

About Bias

Intent

In this activity, students continue to apply combinatorial coefficients.

Mathematics

About Bias requires students to use combinatorial coefficients to evaluate the probability of the null hypothesis that a group of all adults was chosen randomly from a pool containing both adults and teenagers.

Progression

This activity uses combinatorial coefficients in a new application. Students should recall null hypotheses from the unit *Is There Really a Difference* in IMP Year 2.

Approximate Time

5 minutes for introduction

25 minutes for activity

10 minutes for discussion

Classroom Organization

Individuals, followed by whole-class discussion

Doing the Activity

This activity asks students to make a statistical judgment based on probability. They are likely to use combinatorial coefficients to compute the relevant probability.

In introducing the activity, point out that the problem focuses on finding the probability according to a particular assumption or model. It may be helpful to describe this assumption—that the subcommittee was chosen at random—as the null hypothesis of the students who are speaking to the school board.

If groups need a hint getting started, you can ask them to start by finding the number of possible distinct six-person subcommittees.

Discussing and Debriefing the Activity

Have one or two students present their groups' analysis of the probability. There are several ways in which students might find this probability. Here are two possible analyses:

- There are $_{15}C_6$ possible subcommittees. Of these, $_{10}C_6$ consist entirely of adults. Therefore, the probability of picking an all-adult committee is the fraction

$$\frac{_{10}C_6}{_{15}C_6}$$

- If the principal picked people one at a time, the probability that the first person chosen would be an adult is $\frac{10}{15}$. If the first person was an adult, then the probability that the second would also be an adult is $\frac{9}{14}$ (because one adult was already on the subcommittee), and so on. Therefore, the probability of getting an all-adult committee is the product

$$\frac{10}{15} \cdot \frac{9}{14} \cdot \frac{8}{13} \cdot \frac{7}{12} \cdot \frac{6}{11} \cdot \frac{5}{10}$$

Though both of these methods are legitimate, be sure that the first approach is included in the discussion, as a way of illustrating the use of combinatorial coefficients.

In any case, have students give the numerical value of the probability (using graphing calculators to get the combinatorial coefficients if needed). Because $_{10}C_6$ = 210 and $_{15}C_6$ = 5005, the result is $\frac{210}{5005}$, which is approximately .04.

If both methods come up, you might have students write out the combinatorial coefficients as quotients of products so that the expression with combinatorial coefficients becomes

$$\frac{\dfrac{10 \cdot 9 \cdot 8 \cdot 7 \cdot 6 \cdot 5}{6 \cdot 5 \cdot 4 \cdot 3 \cdot 2 \cdot 1}}{\dfrac{15 \cdot 14 \cdot 13 \cdot 12 \cdot 11 \cdot 10}{6 \cdot 5 \cdot 4 \cdot 3 \cdot 2 \cdot 1}}$$

This may help students to see why the two approaches give the same result.

Question 2

Once the probability from Question 1 has been established, have students share opinions about whether the principal stacked the committee. Point out that this is a serious charge requiring strong evidence.

You should bring out that if the principal regularly picked subcommittees and truly chose them at random, then about 1 out of every 25 subcommittees he picked would consist entirely of adults.

Binomial Powers

Intent

This activity is the first stage of preparation for the development of the binomial theorem, which will be completed with the discussion of *Binomials and Pascal—Part II*.

Mathematics

In this activity, students use the distributive property to expand squares and cubes of binomial expressions.

Progression

Binomial Powers is preparatory to further activities in which students will see how combinatorial coefficients can be used in writing binomial expansions.

Approximate Time

30 minutes for activity (at home or in class)

10 minutes for students to check their solutions

Classroom Organization

Individuals, followed by an opportunity for students to compare solutions within their groups

Doing the Activity

This activity contains no new mathematical topics and should require little or no discussion.

Discussing and Debriefing the Activity

You might simply have students compare answers on this activity within groups. As you listen to their discussions, you should get a sense of how well students understand and can use the distributive property. Be sure they see that for Question 7, they can simply multiply their answer from Question 2 by $a + b$. (In *Binomials and Pascal—Part I*, they will look at the results of raising $a + b$ to higher powers.)

Questions 8 through 11 provide a general review of algebraic techniques. We recommend that you not spend too much time on these questions.

Don't Stand for It

Intent

In this activity, students analyze a situation involving the binomial distribution and cumulative probability.

Mathematics

Don't Stand for It describes a small-town lunch counter. The only ten workers who go out to lunch each day have two possible places to eat, and students are asked to determine how many stools the owner should provide in order to have less than a 5% chance of running out of stools on any given day. This is another situation in which combinatorial coefficients are used to calculate probabilities. The discussion focuses on the need to consider cumulative probability.

Progression

This activity provides another application of combinatorial coefficients.

Approximate Time

30 minutes for activity

10 minutes for discussion

Classroom Organization

Individuals or small groups, followed by whole-class discussion

Doing the Activity

No specific introduction is needed for this activity. The key element, aside from the use of the combinatorial coefficients, is recognizing that the owner needs to consider *cumulative* probability.

If students need hints getting started, you can ask what the probability is that all ten people will choose to eat at the second lunch spot, and then ask about nine people, and so on. If students have trouble finding the probability for nine people making that choice, you can suggest that they give names (or assign letters) to the ten people and list the possible combinations of who the nine people could be (and similarly for the case of eight people, and so on).

Discussing and Debriefing the Activity

You might organize the discussion by having different students give the probabilities for having ten, nine, or eight people show up at the lunch counter. Students should see these results:

- The probability of ten people showing up is $.5^{10}$, which is approximately .001.

- The probability of nine people showing up is $\binom{10}{9} \cdot .5^9 \cdot .5^1$, which is approximately .010.

- The probability of eight people showing up is $\binom{10}{8} \cdot .5^8 \cdot .5^2$, which is approximately .044.

At this point in the discussion, you might bring out that the last result is still below the owner's 5 percent threshold and have another student present the probability of seven people showing up.

Students should object, noting that if the owner has only seven stools, then the probability of his not having enough is the sum of the probabilities found so far, and the three results (for ten, nine, and eight people) total approximately .055. Because this is more than .05, the owner needs to have eight stools.

If no one objects when you try to go on to the case of seven stools, simply keep going for a while. At some point, interrupt to ask what the question is in the problem, and help students see what they are looking for.

Stop! Don't Walk!

Intent

This activity continues the theme of using combinatorial coefficients to find probabilities and using statistical reasoning to test hypotheses.

Mathematics

In this activity, students use probability and combinatorial coefficients to evaluate a null hypothesis. Like *Don't Stand for It*, it involves cumulative probabilities. The discussion following the activity reminds students that in determining how unusual a result is, one should consider results "that weird or weirder."

Progression

This activity is another application of combinatorial coefficients to find probabilities.

Approximate Time

30 minutes for activity (at home or in class)

10 minutes for discussion

Classroom Organization

Individuals, followed by whole-class discussion

Doing the Activity

Stop! Don't Walk! describes a situation in which an individual believes that a stoplight she must walk through each day on the way to school is red in her direction with a disproportionate frequency. The activity presents the results for two intervals over which she kept track of whether or not the light was red when she arrived at the intersection and asks students to find the probabilities of each of those results occurring if the light is truly set to be red 60 percent of the time.

Students should be able to do this activity individually, with no introduction.

Discussing and Debriefing the Activity

You might introduce the discussion by asking how the concept of a null hypothesis might be used. Bring out that the null hypothesis here is the statement by the Department of Public Works that the light is red 60 percent of the time within each cycle. Patience is trying to convince the DPW that it should reject that null hypothesis.

Question 1

Let a volunteer present Question 1. Students should see that the probability of five red lights is $.6^5$, which is about $.08$. Thus, five red lights in a row is unlikely but not implausible and Patience is probably right to think that this is not enough evidence to convince the Department of Public Works.

Question 2

Finding the relevant probabilities in Question 2 is a bit more complicated. If students think that the probability of exactly 13 red lights out of 15 is $.6^{13} \cdot .4^2$, remind them of their work on *Five for Seven.* For instance, help them see that the probability of getting 13 red lights and then 2 green lights *is* $.6^{13} \cdot .4^2$, but that they need to consider other sequences.

If needed, ask specifically, How many sequences are there with 13 red and 2 green lights? Students should see that the number is given by the combinatorial coefficient $_{15}C_2$, which is equal to 105. Thus, the probability of getting 13 red and 2 green lights is $105 \cdot .6^{13} \cdot .4^2$, which is approximately $.022$.

Why Consider 14 and 15?

Ask, Why does Question 2 ask about the probability of getting 13 'or more' red lights? Bring out that if the null hypothesis should be rejected if Patience gets 13 red lights, it should also be rejected if she gets more than 13. Thus, the answer to Question 2 is the sum

$$_{15}C_2 \cdot .6^{13} \cdot .4^2 + {}_{15}C_1 \cdot .6^{14} \cdot .4^1 + .6^{15}$$

which is approximately $.027$.

Making Decisions

You can let students share their ideas about whether this probability is sufficiently small that the city should conduct a test of the light. Students may also have other ideas about Patience's methodology.

Key Questions

How many sequences are there with 13 red and 2 green lights?

Why does Question 2 ask about the probability of getting 13 'or more' red lights?

Supplemental Activity

My Dog's Smarter Than Yours (reinforcement) requires a strategy similar to that for *Stop! Don't Walk!*

Pascal's Triangle

Intent

In this section, students explore Pascal's triangle and its relationship to combinatorial coefficients.

Mathematics

Pascal's triangle, named for the mathematician Blaise Pascal, is a triangular array of numbers in which both ends of each row are 1's and every other entry is the sum of the two nearest numbers above. The numbers in Pascal's triangle are all combinatorial coefficients, and the numbers in each row are the coefficients of the terms in the expansion of a binomial raised to the power equal to the row number. (In fact, that is why they are called combinatorial *coefficients*.)

The work in this section with Pascal's triangle has several goals:

- To show students another way to find the numerical values of the combinatorial coefficients
- To give students an opportunity to work with the meaning of $_nC_r$ on a more abstract level
- To have students develop the powerful binomial theorem and see some of its consequences
- To improve students' investigative skills

Progression

In *Pascal's Triangle*, students will look for any patterns they can find and prepare presentations. In *Hi There!*, they will look at the classic 'handshake problem' and see that answers to that problem correspond to entries in Pascal's triangle. But they also see that these answers can be described as combinatorial coefficients. They are then told that *all* the numbers in Pascal's triangle are combinatorial coefficients.

In *Pascal and the Coefficients,* students begin their exploration of the connection between Pascal's triangle and the combinatorial coefficients. This exploration continues with *Combinations, Pascal's Way,* in which students seek to explain various features of Pascal's triangle in terms of the meaning of combinatorial coefficients. These features include the symmetry of the array and the fact that the sum of the entries in each row is a power of 2.

Binomials and Pascal—Part I continues the exploration of the coefficients in the expansion of $(a + b)^n$ begun in *Binomial Powers.* Students see that the coefficients are entries in Pascal's triangle and apply this fact in *Binomials and Pascal—Part II*. They learn that this relationship is called the **binomial theorem** and examine *why* combinatorial coefficients appear in the expansion of $(a + b)^n$. In *A Pascal Portfolio*, they summarize their work for the investigation.

Pascal's Triangle

Hi There!

Pascal and the Coefficients

Combinations, Pascal's Way

Binomials and Pascal—Part I

Binomials and Pascal—Part II

A Pascal Portfolio

Pascal's Triangle

Intent

In this activity, students identify patterns and properties of Pascal's triangle.

Mathematics

Pascal's triangle is a useful tool for generating combinatorial coefficients. One way to define Pascal's triangle is by a rule for generating each row from the previous one. Another way is to define each entry of each row as an appropriate combinatorial coefficient. In *Pascal's Triangle*, students are introduced to Pascal's triangle and look for any patterns they can find.

Progression

Over the next several activities, students will see that the numbers in Pascal's triangle are combinatorial coefficients, then explain some of the patterns in the triangle through the meaning of combinatorial coefficients. They will then find that the coefficients of terms in the expansion of a power of a binomial are entries from Pascal's triangle, and thus combinatorial coefficients.

Approximate Time

30 minutes for activity

30 minutes for presentations and discussion

Classroom Organization

Groups, followed by group presentations and whole-class discussion

Materials

Poster materials for each group

Doing the Activity

To get students started on today's activity, give each group some chart paper on which they can write their observations. Tell them that they will be presenting some of their observations and conclusions to the class.

Discussing and Debriefing the Activity

As observations are presented, make a master list that will be posted in class, because students will be working further with these observations in *Combinations, Pascal's Way*.

Begin by asking one or two groups to present the patterns their groups used to extend Pascal's triangle to more rows. Here are two common ways they may express the pattern:

- Each entry is the sum of the two closest entries in the previous row.
- The sum of two adjacent entries in a given row is equal to the entry between them in the next row.

Be sure everyone understands how to use a method of this type to get additional rows of Pascal's triangle. Tell the class that this approach is one way of *defining* Pascal's triangle.

Have volunteers give additional rows for Pascal's triangle, so that you have at least ten complete rows altogether. Post this extended version of Pascal's triangle for easy reference for the rest of the unit.

Other Observations

After this main pattern is clear, have other groups present other observations about Pascal's triangle. Encourage students to share ideas even if they think their observations are "obvious."

As groups present, be sure that these ideas are mentioned in the presentations:

- The arrangement of numbers is symmetrical.
- Each row begins and ends with the number 1.
- From the second row on, the second entry in the row is one less than the "row number."

Here are two other patterns students might find:

- The sum of the entries for the nth row is 2^{n-1}.
- Each row has its largest entry (or largest two entries) in the middle of the row.

(In the discussion of the next activity, *Hi There!*, you'll introduce the idea of labeling the top row as "row 0" and the first entry in each row as "entry 0," but here students will natural call them "row 1" and "entry 1".)

Note: It's unlikely at this stage that students will recognize the numbers in Pascal's triangle as combinatorial coefficients. If they do, tell them that this will be addressed soon.

Hi There!

Intent

In this activity, students see how combinatorial coefficients relate to a classic problem.

Mathematics

Hi There! introduces students to the classic handshake problem: how many handshakes take place if n people are in a room and everyone shakes hands exactly once with everyone else. Students discover that they can express the solution to the problem in terms of combinatorial coefficients, and that Pascal's triangle is also related.

The discussion reveals that all of the entries in Pascal's triangle are combinatorial coefficients and describes the standard labeling system, in which the first row is "row 0" and the first entry in each row is "entry 0."

Progression

After a demonstration, students work on the activity individually and then discuss their results as a class. The discussion also introduces the standard labeling system for Pascal's triangle.

Approximate Time

10 to 20 minutes for introduction

30 minutes for activity (at home or in class)

20 minutes for discussion

Classroom Organization

Individuals, followed by whole-class discussion

Doing the Activity

Start by having the class act out a simple example of the handshake problem to be sure students understand it. For instance, have four students come up and each shake one another's hand, so students see that in this case, six handshakes are needed.

Discussing and Debriefing the Activity

Students should be able to put together an In-Out table of values like this:

Number of people	Number of handshakes
2	1
3	3
4	6
5	10

Ask, **What pattern do you see in the 'Out' numbers?** They may have seen that each *Out* can be obtained by adding the previous *In* to the previous *Out*. (For instance, 10 can be written as 6 + 4.) Or they may have recognized the values in the second column as the **triangular numbers:** 1, 1 + 2, 1 + 2 + 3, 1 + 2 + 3 + 4, and so on. (If they notice this pattern, you can introduce the term *triangular numbers* and ask if anyone can explain why this pattern of numbers occurs in the problem.)

Comment: These are called *triangular numbers* because they represent the number of dots in triangular arrangements like these:

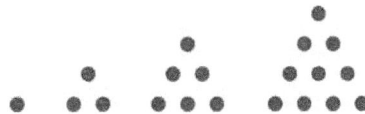

Ask, **Did anyone get a general formula for the case of *n* people shaking hands?** You may get more than one possibility. We suggest that you verify each suggestion with one or two specific cases but not ask for explanations until the discussion of Question 4. (If you do not get any suggestions for the general formula, you can leave that as well until the discussion of Question 4.)

Question 3

Next, ask, **How are the 'Out' values related to Pascal's triangle?** If needed, help the class to see that these numbers appear along one of the diagonals of Pascal's triangle, as shown by the boxed entries here:

```
              1
           1     1
        1     2     1
     1     3     3     1
  1     4     6     4     1
1     5    10    10     5     1
1  6  [15]  20   15    6    1
```

Question 4

Ask, **How is this problem related to combinatorial coefficients?** As a hint, ask, **How many of the *n* people are involved in each handshake?** (two) Then ask how the number of *pairs of people* can be expressed using a combinatorial coefficient. (Or you can suggest that students think of the *n* people as like *n* flavors of ice cream and see each handshake as like a two-scoop bowl of ice cream.)

Students should see that the appropriate coefficient is $_nC_2$. Have a volunteer give a formula for $_nC_2$.

If needed, review how $_nC_2$ is related to $_nP_2$. The class should see that

$_nP_2 = n(n-1)$ and that $_nC_2 = \frac{_nP_2}{2}$. Together, these facts give $_nC_2 = \frac{n(n-1)}{2}$.

Have students verify that the formula for $_nC_2$ fits their table of data. They should also compare this formula with any formulas they found for Question 2 and verify that the expressions are equivalent (assuming that the earlier results were correct).

Pascal's Triangle and Combinatorial Coefficients

Tell the class that it is no accident that the answers to the handshake problem appear both in Pascal's triangle and as combinatorial coefficients and that, in fact, Pascal's triangle is made up entirely of combinatorial coefficients.

You might let students try to figure out what the relationship is—that is, ask, **How are *n* and *r* of *nCr* related to the row number and entry number of a number in Pascal's triangle?** You might point out that *n* and *r* can both start at 0 rather than at 1.

Bring out that to get the numbering to match up properly, we generally refer to the first line of Pascal's triangle as "row 0" and the first entry in each row as "entry 0." Thus, in row *n* [which is actually the $(n + 1)^{th}$ row], the entries go from entry 0 through entry *n*. Tell students that based on this numbering system, entry *r* of row *n* of Pascal's triangle is $_nC_r$.

To clarify this system, ask, **What are *n* and *r* for the 1 at the top of the triangle?** Bring out that this 1 represents $_0C_0$. Identify another initial entry in a row and have students find *n* and *r* for that entry as well. (They might realize right away that *r* = 0, but they may need to stop and think about the value of *n*.)

Ask, **Where should $_nC_2$ appear in Pascal's triangle?** ($_nC_2$ is from the handshake problem.) Students should see that it should actually be the third entry in the $(n + 1)^{th}$ row (using the initial counting system). Have them confirm that this is correct. For instance, the boxed number 10 shown on the preceding page is equal to $_5C_2$, and this appears as the third number in the sixth row of Pascal's triangle. This example helps explain why we begin numbering from 0, so that $_5C_2$ is more appropriately described as entry 2 of row 5.

Key Questions

What pattern do you see in the 'Out' numbers?

Did anyone get a general formula for the case of n people shaking hands?

How are the 'Out' values related to Pascal's triangle?

How is this problem related to combinatorial coefficients?

How many of the n people are involved in each handshake?

How are *n* and *r* of $_nC_r$ related to the row number and entry number of a number in Pascal's triangle?

What are *n* and *r* for the 1 at the top of the triangle?

Where should $_nC_2$ appear in Pascal's triangle?

Pascal and the Coefficients

Intent

In this activity, students use the fact that the entries in Pascal's triangle are combinatorial coefficients.

Mathematics

This activity clarifies the connection between Pascal's triangle and the binomial coefficients. Students verify that the entries in Pascal's triangle appear to be combinatorial coefficients and use that fact as a shortcut in evaluating coefficients in the form $\binom{n}{r}$. They also see once more that $\binom{n}{0}$ and $\binom{n}{n}$ are both equal to 1 and explain this in terms of the ice cream analogy.

Progression

Students work on the activity individually and then discuss their results as a class.

Approximate Time

25 minutes for activity (at home or in class)

10 minutes for discussion

Classroom Organization

Individuals, followed by whole-class discussion

Doing the Activity

Students should be able to complete this activity individually with no introduction.

Discussing and Debriefing the Activity

Question 1 provides an opportunity to review the meaning of combinatorial coefficients very concretely. In addition to the "bowls of ice cream" model, elicit other methods for computing them. (For instance, students might use the formulas they developed in *Formulas for $_nP_r$ and $_nC_r$.*)

The goal of Question 2 is simply to verify that students know how to relate specific combinatorial coefficients to their positions within Pascal's triangle. Have students locate these combinatorial coefficients on the class poster version of Pascal's triangle.

Question 3

Question 3 can serve as a model for the next activity, *Combinations, Pascal's Way,* so it's important to take time to discuss this problem.

Before getting into explanations, be sure students see that the statement "$\binom{n}{0}$ and $\binom{n}{n}$ are both equal to 1" expresses the fact that each row begins and ends with the number 1.

Then have them explain why this is so, which they might do using the idea of a zero-scoop bowl of ice cream or with some other model. Bring out that verifying that the expression $_nC_0$ is always 1 is a partial step toward showing that the entries in Pascal's triangle are actually the combinatorial coefficients.

Combinations, Pascal's Way

Intent

In this activity, students explain patterns in Pascal's triangle in terms of the meaning of combinatorial coefficients.

Mathematics

Combinations, Pascal's Way improves students' investigative skills while strengthening their understanding of combinatorial coefficients by requiring them to work with them on a more abstract level.

Progression

In the last question of *Pascal and the Coefficients*, students used the ice cream analogy for combinatorial coefficients to explain one pattern that can be observed in Pascal's triangle. Now they are asked to express each of the patterns they found in *Pascal's Triangle* using combinatorial coefficients and explain each pattern using the meaning of the combinatorial coefficients.

Approximate Time

35 to 40 minutes for activity

40 minutes for discussion

Classroom Organization

Small groups, followed by group presentations and whole-class discussion

Doing the Activity

Review the master list of patterns and relationships within Pascal's triangle made following *Pascal's Triangle*. The list should include at least these items:

- Each entry is the sum of the two closest entries in the previous row.
- The arrangement of numbers is symmetrical.
- Entry 1 in row n is equal to n.

(The property that each row begins and ends with 1 was discussed in *Pascal and the Coefficients*.)

Here are some other ideas that might have been part of the class list:

- The sum of the entries in row n is 2^n.
- Each row has its largest entry (or largest two entries) in the middle.

Note: Some of these properties were stated slightly differently in the discussion of *Pascal's Triangle*, because that discussion preceded the introduction of the idea that the first row is called "row 0" and that the first entry in each row is called "entry 0."

Setting Priorities

If students seem to be having trouble getting started on the activity, we recommend that you have them choose one of the properties from the list and focus initially simply on stating the property in terms of combinatorial coefficients. Use Question 3 of *Pascal and the Coefficients* as a model of how to proceed.

Each of the properties involves its own particular issues, and some are easier to deal with than others. Here we discuss the properties in a sequence that may be most productive for your students to work on. The discussion ideas under *Discussing and Debriefing the Activity* give further details, in case you want hints for students.

The arrangement of numbers is symmetrical: The symmetry property may be a good starting point, because students have seen this aspect of combinatorial coefficients in several contexts. Stating the symmetry of Pascal's triangle as the equation $_nC_r = {_nC_{n-r}}$ is an important first step.

Be sure at least one group formulates this generalization. As a hint, you can ask what entry is situated symmetrically to $_{50}C_{13}$. Further hints can build on the discussion that grew out of Question 3 of *What's for Dinner?*

Entry 1 in row *n* is equal to *n*: This is similar to Question 3 from *Pascal and the Coefficients* and should be fairly accessible to students.

Each entry is the sum of entries from the previous row: This is essentially the defining property of the pattern. That is, it is the property that allows us to generate more rows. Because of this, the connection between this property and the combinatorial coefficients is especially important. (Showing that the combinatorial coefficients have this property is the induction step in a general proof that entry *r* of row *n* is always $_nC_r$.)

Stating and proving this property is fairly challenging, which is why we do not recommend it as a first step in the activity. If the class does not seem ready to tackle this, you can leave it unresolved.

If any groups seem ready to tackle this property, you can begin by having them look at a specific example. For instance, ask them to use combinatorial coefficients to express the fact that the boxed entry 15, in the array below, is the sum of the entries 5 and 10 in the previous row.

$$
\begin{array}{ccccccccccccc}
& & & & & & 1 & & & & & & \\
& & & & & 1 & & 1 & & & & & \\
& & & & 1 & & 2 & & 1 & & & & \\
& & & 1 & & 3 & & 3 & & 1 & & & \\
& & 1 & & 4 & & 6 & & 4 & & 1 & & \\
& 1 & & 5 & & 10 & & 10 & & 5 & & 1 & \\
1 & & 6 & & \boxed{15} & & 20 & & 15 & & 6 & & 1
\end{array}
$$

They should see that this can be expressed by the equation $_6C_2 = {}_5C_1 + {}_5C_2$. They can work from examples like this to formulate a generalization and then consider why such a property should hold.

The row sum is a power of 2: You might ask students to state this property using summation notation. Proving this property is challenging, but students may find good explanations. (One nice proof uses the binomial theorem.) Students continue their work on developing this theorem in *Binomials and Pascal—Part I*.

The largest entries are in the middle: Stating this clearly and proving it are quite challenging. You might have students begin with specific examples. For instance, they might try to show, using the formula for $_nC_r$, that $_{100}C_{50}$ is the largest combinatorial coefficient of the form $_{100}C_r$. (This general property is the subject of the supplemental activity *Maximum in the Middle*.)

Discussing and Debriefing the Activity

Let groups take turns making presentations on different properties of Pascal's triangle. As each pattern or relationship is discussed, let other groups add further insights or explanations.

We provide here some further ideas about each of the properties discussed in the section *Doing the Activity* above. As indicated there, students need not formulate or prove all of the properties, and these ideas are included primarily as background material for you.

The arrangement of numbers is symmetrical: If necessary, have presenters begin with specific examples. Once the class has formulated the general equation $_nC_r = {}_nC_{n-r}$, focus on explanations for the symmetry based on the meaning of combinatorial coefficients. Students can build on a variety of situations for this. For example, in *Top That Pizza!*, they saw that there were the same number of pizzas using three of the five toppings as there were using two of the five. Similarly, in *Five for Seven*, they saw that there were the same number of sequences with two wins out of seven as there were for five wins out of seven.

Students should be able to articulate that any time you choose a set of r objects out of n, you have indirectly also chosen $n - r$ objects to be excluded from the set.

Entry 1 in row n is equal to n: There may be some confusion about this property because the entry in question is actually the *second* entry of the row. Once students sort this out, they should be able to state the property as the equation $_nC_1 = n$ and explain this using one of the models for $_nC_r$. For instance, they should see that Jonathan can create exactly n distinct one-scoop bowls of ice cream.

As a follow-up, you might ask students to formulate a principle symmetric to this one. That is, try to elicit the general equation $_nC_{n-1} = n$.

Each entry is the sum of entries from the previous row: As was suggested previously, students might use specific examples, such as $_6C_2 = {}_5C_1 + {}_5C_2$, to develop a generalization. The generalization can take different forms, depending on which term in the equation is represented by the general expression $_nC_r$. For instance, either of these equations could be used:

- $_nC_r = {}_{n-1}C_{r-1} + {}_{n-1}C_r$
- $_{n+1}C_{r+1} = {}_nC_r + {}_nC_{r+1}$

Here is one possible general explanation of this pattern, using the first form of the equation and building on the bowls-of-ice-cream metaphor. (Students may find it more helpful to use the same reasoning with specific numbers for n and r.)

We can think of $_nC_r$ as the number of different bowls of ice cream that Jonathan can create if he goes into an ice cream store with n flavors and wants a bowl with r scoops (of different flavors).

Suppose one of the flavors the store offers is vanilla. Then some of the possible bowls of ice cream include vanilla, and some do not. We can express the number of bowls of each type using combinatorial coefficients.

Jonathan needs r scoops altogether. If one of them is vanilla, then he needs to choose $r - 1$ more scoops. He's already used vanilla, so these $r - 1$ scoops must come from the remaining $n - 1$ flavors. Thus, there are $_{n-1}C_{r-1}$ distinct r-scoop bowls of ice cream that include vanilla as one of the flavors.

On the other hand, if he does not use vanilla, then he needs to choose all r scoops from among the other $n - 1$ flavors. Thus, there are $_{n-1}C_r$ distinct r-scoop bowls of ice cream that do not include vanilla.

Because the total number of different bowls is the sum of the number of ways that include vanilla plus the number of ways that don't include vanilla, we have

$$_nC_r = {_{n-1}C_{r-1}} + {_{n-1}C_r}$$

The row sum is a power of 2: Here, as with some other properties, the numbering of the rows may be problematic. Based on how we have labeled Pascal's triangle, the correct statement is that the sum of the entries in row n is 2^n. For instance, the sum $_4C_0 + {_4C_1} + {_4C_2} + {_4C_3} + {_4C_4}$ is equal to 2^4.

Here are a couple of ways students might explain this property:

- *Using the baseball situation:* Each term in the sum

$$_7C_0 + {_7C_1} + \ldots + {_7C_6} + {_7C_7}$$

gives the number of win/loss sequences in which the Good Guys win a particular number of games. The sum is the total number of possible sequences of outcomes. Because each game has two possible outcomes, the number of possible sequences of outcomes is 2^7. (In more abstract terms, this property states that a set with n elements has 2^n subsets. Each combinatorial coefficient gives the number of subsets of a particular size.)

- *Using the defining pattern:* In creating each new row of Pascal's triangle, each number of the preceding row is used twice (namely, in finding the number below it to the right and in finding the number below it to the left; this includes the 1's on the ends). Therefore, each row sum is twice the preceding row sum. Row 0 has a row sum of 2^0, row 1 has a row sum of 2^1, and the exponent goes up by 1 along with the row number.

The largest entries are in the middle: This property is difficult to formulate because it has two cases, depending on whether the number of entries in the row is odd or even.

One approach to explaining the pattern is to look at the ratio of a given entry to the preceding entry—that is, the ratio $\dfrac{_nC_r}{_nC_{r-1}}$ —and examine what values of r make this ratio greater than 1. Students can use the formula

$$_nC_r = \frac{n!}{r! \cdot (n-r)!} \cdot$$

substituting this and a similar expression for $_nC_{r-1}$ in the ratio $\dfrac{_nC_r}{_nC_{r-1}}$. They can then simplify the overall expression, examining the simplified expression to determine what relationship between n and r will make $_nC_r$ greater than, equal to, or less than $_nC_{r-1}$.

Supplemental Activities

Defining Pascal (extension) challenges students to explain the defining pattern of Pascal's triangle using combinatorial coefficients, if they did not do so as part of this activity.

Maximum in the Middle (extension) asks students to use the formulas for combinatorial coefficients to prove that the center entry (or entries) of each row in Pascal's triangle will be the largest value(s) in that row, again if they did not do so as part of this activity.

Binomials and Pascal—Part I

Intent

This activity will lead to the development of the binomial theorem.

Mathematics

In this activity, students expand $(a + b)$ to the power of 2, 3, 4, and 5, then recognize the connection between entries in Pascal's triangle and coefficients in powers of binomials. For example, when you expand and simplify $(a + b)^5$, the coefficients of each term are the entries of row 5 in Pascal's triangle.

Progression

This activity begins an exploration of ideas that will continue in *Binomials and Pascal—Part II*.

Approximate Time

30 minutes for activity (at home or in class)

10 minutes for discussion

Classroom Organization

Individuals, followed by whole-class discussion

Doing the Activity

Some students may need support with the algebra of this activity.

Discussing and Debriefing the Activity

Begin by having volunteers give you the expansions of $(a + b)^2$, $(a + b)^3$, $(a + b)^4$, and $(a + b)^5$. If the expansions are not organized in standard order (starting with the highest power of a and with terms combined in descending powers of a), then get students to arrange them that way.

Because students have been examining Pascal's triangle for the last several days, they will probably recognize that the coefficients are the entries of the appropriate row of that array.

Bring out that every term of these expansions can be written using both a power of a and a power of b. For instance, we can write the initial term of $(a + b)^5$ as $a^5 b^0$ (rather than simply as a^5), and we can write the term $5a^4 b$ as $5a^4 b^1$. Seeing this may make the pattern of exponents clearer for some students.

Follow up by asking, What might a term in the expansion of $(a + b)^{20}$ look like? If needed, you might ask specifically, What power of b will go with a^{16}? What is the coefficient for this term? They should see that this term has a companion factor of b^4 and that the coefficient is $_{20}C_{16}$. (They need not find the numerical value of $_{20}C_{16}$.)

Students will continue their exploration of this idea in *Binomials and Pascal–Part II* and they will then examine why combinatorial coefficients occur in the expansion of these binomials.

Key Questions

What might a term in the expansion of $(a + b)^{20}$ look like?

What power of b will go with a^{16}? What is the coefficient for this term?

Binomials and Pascal—Part II

Intent

This activity concludes students' work in this unit with the binomial theorem and Pascal's triangle.

Mathematics

Binomials and Pascal—Part I. asks students to expand several binomials using Pascal's triangle.

In the discussion following this activity, students state the binomial theorem in general form. Then, students use a baseball interpretation (from the central unit problem) of a particular binomial expansion to help explain why the expansion involves combinatorial coefficients. They also use the binomial theorem to show indirectly that the probabilities in the unit problem add up to 1.

Progression

This activity applies and extends the ideas of *Binomials and Pascal—Part I.*

Approximate Time

30 minutes for activity (at home or in class)

45 to 50 minutes for discussion

Classroom Organization

Individuals, followed by whole-class discussion

Doing the Activity

This activity applies the relationships that students learned in *Binomials and Pascal—Part I.* They should be able to do this activity with no introduction.

Discussing and Debriefing the Activity

Let a volunteer present each question. On Question 1, be sure students see that they can simply write down the pattern of terms and use Pascal's triangle to get the coefficients. That is, the terms will involve the expressions a^{10}, a^9b, a^8b^2, and so on, ending with ab^9 and b^{10}, with corresponding coefficients from row 10 of Pascal's triangle.

On Question 2, bring out that students can apply Pascal's triangle to get

$$(a + b)^5 = a^5 + 5a^4b + 10a^3b^2 + 10a^2b^3 + 5ab^4 + b^5$$

and then simply replace b with 2 and do the appropriate simplification. For instance, the term $10a^2b^3$ becomes $10a^2 \cdot 2^3$, which is $80a^2$. Point out that this approach is probably easier than applying the distributive property repeatedly to expand $(a + 2)^5$.

On Question 3, the main focus should be on the signs of the terms. Bring out that if the terms are written in descending powers of x, then the signs will alternate because powers of -1 alternate between 1 and -1. In absolute value, the coefficients for the expansion of $(x - 1)^4$ are the entries of row 4 of Pascal's triangle.

The Binomial Theorem

Tell students that the expansions they have seen are examples of an important general principle called the **binomial theorem** and that this theorem is the reason the combinatorial coefficients are also called **binomial coefficients.** (In fact, this also explains the *coefficient* part of the term *combinatorial coefficient*.)

Ask, **How might you write the general form of this theorem?**, and let them work on this in groups if needed. As a clarification, you can ask how they might write the general term of the expansion of the expression $(a + b)^n$. You might suggest that they use summation notation to write the overall expansion.

After a few minutes, let several students describe their methods for stating the binomial theorem. Here are three possible ways students might write it. ($\binom{n}{r}$ is used here, but students might use $_nC_r$ instead.)

- Using numerical values for the first two and last two terms, but expressing the middle terms using $\binom{n}{r}$ notation:

$$a^n + na^{n-1}b + \binom{n}{2}a^{n-2}b^2 + \binom{n}{3}a^{n-3}b^3 + K + nab^{n-1} + b^n$$

- Writing all coefficients using $\binom{n}{r}$ notation:

$$\binom{n}{0}a^n + \binom{n}{1}a^{n-1}b + \binom{n}{2}a^{n-2}b^2 + \binom{n}{3}a^{n-3}b^3 + K + \binom{n}{n-1}ab^{n-1} + \binom{n}{n}b^n$$

- Using summation notation:

$$\sum_{r=0}^{n} \binom{n}{r} a^{n-r} b^r$$

The Binomial Theorem and Baseball

Write out this product for students to see, and ask them to imagine each W or L as representing a win or loss for the Good Guys.

$$(W_1 + L_1)(W_2 + L_2)(W_3 + L_3)(W_4 + L_4)(W_5 + L_5)(W_6 + L_6)(W_7 + L_7)$$

Then have them imagine multiplying the expression out. They need not actually do the expansion, but bring out that the expansion would be a sum of many terms, and that each term of the expansion would be a sequence of W's or L's. You might have them give a sample term, such as $W_1 L_2 W_3 L_4 W_5 L_6 W_7$, emphasizing that each term of the expansion contains one factor from each factor of the original product.

Next, ask, **What might a term of the expansion mean in terms of the baseball situation?** For instance, the term $W_1 L_2 W_3 L_4 W_5 L_6 W_7$ represents the outcome in which the Good Guys win games 1, 3, 5, and 7 and lose games 2, 4, and 6.

Then ask, **What does such a product have to do with the binomial theorem and combinatorial coefficients?** Be sure students see, to begin with, that if they omitted the subscripts, the expression would be $(W + L)^7$.

If needed, ask specifically, **How can you use the baseball interpretation of the product to explain the binomial theorem?**—in particular, the fact that the coefficients of $(a + b)^n$ are combinatorial coefficients. Bring out that if they think of W's and L's as wins and losses, then terms that become alike when the subscripts are dropped also represent the same overall result for the Good Guys. For instance, the terms $W_1 L_2 W_3 L_4 W_5 L_6 W_7$ and $L_1 L_2 W_3 W_4 W_5 L_6 W_7$ would both become $W^4 L^3$ and would both represent sequences in which the Good Guys win four of their games and lose three.

Bring out that the number of such sequences is the number of ways in which the Good Guys get this result, which is the combinatorial coefficient $_7C_4$. On the other hand, by dropping subscripts, this becomes the coefficient of the term $W^4 L^3$. In other words, the coefficient of the term $W^4 L^3$ in the expansion of $(W + L)^7$ is the same as the number of ways in which the Good Guys can win four of their games and lose three, which is the combinatorial coefficient $_7C_4$.

Another approach to this discussion is to create a seven-stage tree diagram, in which each stage involves the choice between a W or an L. You can bring out that each path through this tree diagram is analogous to a term of the expansion of $(W + L)^7$, which consists of products involving either a W or an L from each factor.

The Binomial Theorem and Probabilities

You can use the preceding discussion to help students gain further insight into the probabilities involved in the unit problem. Ask, **What happens if you substitute probabilities for the W's and L's?** That is, substitute .62 for W and .38 for L after multiplying out the product

$$(W_1 + L_1)(W_2 + L_2)(W_3 + L_3)(W_4 + L_4)(W_5 + L_5)(W_6 + L_6)(W_7 + L_7)$$

You might have students focus on a particular term or set of like terms. For instance, the term $W_1W_2L_3W_4W_5W_6L_7$ becomes $.62^5 \cdot .38^2$, which is the probability that the Good Guys will win games 1, 2, 4, 5, and 6 and lose games 3 and 7. Combining this with the fact that there are $\binom{7}{5}$ terms with five W's and two L's corresponds to the fact that the probability that the Good Guys will win exactly five of their seven games is $\binom{7}{5} \cdot .62^5 \cdot .38^2$.

Then ask, **What must all the probabilities add up to?** Although students should recognize that the sum of all the probabilities is always 1, help them to see that the binomial theorem shows this another way, because substituting .62 for the W's and .38 for the L's in the product, *before multiplying out,* simply gives $(.62 + .38)^7$ which equals 1. On the other hand, when they multiply out, combine like terms, and substitute the values, they get the probabilities for the different possible outcomes. In other words, multiplying out essentially yields the equation

$$\left(.62 + .38\right)^7 = \sum_{r=0}^{7} \binom{7}{r} \cdot .62^r \cdot .38^{7-r}$$

where the expression on the right is the sum of the probabilities for the different possible outcomes.

Key Questions

How might you write the general form of this theorem?

What might a term of the expansion mean in terms of the baseball situation?

What does such a product have to do with the binomial theorem and combinatorial coefficients?

How can you use the baseball interpretation of the product to explain the binomial theorem?

What happens if you substitute probabilities for the W's and L's?

What must all the probabilities add up to?

Supplemental Activities

The Why's of Binomial Expansion (extension) uses an approach similar to that described at the end of the subsection *The Binomial Theorem and Baseball* to develop a general proof that binomial expansions involve combinatorial coefficients.

The Binomial Theorem and Row Sums (extension) asks students to use the binomial theorem to prove that the row sums in Pascal's triangle are powers of 2, in a manner similar to the reasoning in the subsection *The Binomial Theorem and Probabilities*.

A Pascal Portfolio

Intent

This activity will be part of the unit portfolio, in which students reflect on the main ideas from the unit.

Mathematics

In this activity, students summarize what they have learned about Pascal's triangle.

Progression

A Pascal Portfolio asks students to write a summary of what they know about Pascal's triangle. This activity needs to be saved for inclusion in the unit portfolio.

Approximate Time

30 minutes for activity (at home or in class)

0 to 10 minutes for discussion (optional)

Classroom Organization

Individuals, followed by optional classroom discussion

Doing the Activity

Students should do this activity independently.

Discussing and Debriefing the Activity

You may want to simply collect this activity to assess what students have learned, or you may also want to have volunteers share parts of their work now as a way of reviewing key ideas.

The Baseball Finale

Intent

In this section, students conclude and summarize the unit.

Mathematics

Students synthesize what they have learned about probability and combinatorial coefficients to complete the solution of the central unit problem. They also conclude their study of combinatorial coefficients from the previous section with a brief introduction to binomial distributions. The entire unit is summarized in the unit portfolio and assessments.

Progression

In *Race for the Pennant! Revisited* students solve the central unit problem. The situation from that problem is used to introduce binomial distributions in *Graphing the Games*. In *Binomial Probabilities*, students develop a formula for the probabilities associated with binomial distributions. The unit concludes with *Pennant Fever Portfolio*.

Race for the Pennant! Revisited

Graphing the Games

Binomial Probabilities

Pennant Fever Portfolio

Race for the Pennant! Revisited

Intent

In this activity, students put ideas together to solve the unit problem.

Mathematics

There are no new mathematical ideas in this activity, but students put together what they have learned about probability and combinatorial coefficients to solve the central unit problem.

Progression

Students now return to the unit problem, after side trips to explore probability problems involving combinatorial coefficients in *Combinatorial Reasoning* and Pascal's triangle and the binomial theorem in *Pascal's Triangle*. They have actually had the knowledge necessary to complete the problem since they were not much more than halfway through the unit.

After solving the problem, the class can compare the results with the estimates they made in *Race for the Pennant!* at the beginning of the unit.

Approximate Time

45 minutes for activity

25 minutes for discussion

Classroom Organization

Groups, followed by whole-class discussion

Doing the Activity

Students have known the necessary concepts and techniques to complete the unit problem since at least *More Five for Sevens*. Some may already have done the work asked for in this activity. If so, they can work on their write-ups.

You may want to introduce this final activity by having a whole-class review of the status of work on the unit problem. Have students take out the charts they began with *Possible Outcomes* and *How Likely Is All Wins?* About a third of the cells should already have probabilities filled in (see *More Five for Sevens*).

You may want to suggest to students that as a next step, they find the remaining probabilities for the entries at the top and left of the chart (that is, the probability for each team of winning exactly three or exactly four games out of the remaining seven).

Note: The activity does not explicitly ask students to find the probabilities for all the cells in their charts, and some students might find the probability that the Good Guys will win the pennant without finding the probability for every cell. Thus, if you want students to complete the chart, you will need to say so.

Discussing and Debriefing the Activity

The charts provide an excellent vehicle to allow many students to participate in the final discussion. One approach is to begin by having students from different groups give and explain the probabilities at the top and left (that is, the probabilities for each team individually).

You can then go around the room having individual students give the probability for a single cell within the chart. Finally, have volunteers give the overall probabilities for the Good Guys winning the pennant or tying with the Bad Guys.

Here is the completed chart, showing the probabilities to the nearest .0001 (except for cases in which the probability is less than .0001):

Good Guys' record for the final seven games

		7–0 p = .0352	6–1 p = .1511	5–2 p = .2778	4–3 p = .2838	3–4 p = .1739	2–5 p = .0640	1–6 p = .0131	0–7 p = .0011
Bad	7–0 p = .0280	G .0010	G .0042	G .0078	T .0079	B .0049	B .0018	B .0004	B .00003
Guys'	6–1 p = .1306	G .0046	G .0197	G .0363	G .0371	T .0227	B .0084	B .0017	B .0001
record	5–2 p = .2613	G .0092	G .0395	G .0726	G .0741	G .0454	T .0167	B .0034	B .0003
for	4–3 p = .2903	G .0102	G .0439	G .0806	G .0824	G .0505	G .0186	T .0038	B .0003
the	3–4 p = .1935	G .0068	G .0292	G .0538	G .0549	G .0337	G .0124	G .0025	T .0002
final	2–5 p = .0774	G .0027	G .0117	G .0215	G .0220	G .0135	G .0050	G .0010	G .00009
seven	1–6 p = .0172	G .0006	G .0026	G .0048	G .0049	G .0030	G .0011	G .0002	G .00002
games	0–7 p = .0016	G .00006	G .0002	G .0005	G .0005	G .0003	G .0001	G .00002	G .000002

Adding the probabilities for all the cells labeled G gives the probability of the Good Guys winning as approximately .927, or roughly 93%. Similarly, the probability of a tie is approximately .051, and the probability of the Bad Guys winning is approximately .021. (Due to rounding, these approximate values do not add up to exactly 1.)

If you figure that in the case of a tie, there will be a playoff game between the two teams and that the Good Guys have about a 50 percent chance of winning this game, then their probability of ending up as champions becomes about

$.927 + \frac{1}{2}(.051)$, or about .95. Pretty good!

At the beginning of the unit (in *Race for the Pennant!*), students were asked to state what outcome they thought was most likely for each team and to estimate the probability of the Good Guys winning the pennant. You can now have them compare those conjectures to these actual results.

Simplifications in the Computations

Ask, **Do you have any shortcuts for finding the probability of the Good Guys winning?** For instance, as noted in the discussion following *How Likely Is All Wins?*, if the Good Guys win all their remaining games, then it doesn't matter what the Bad Guys do. This is reflected in the fact that the entire first column of the chart is labeled G. That is, if all you care about is the Good Guys' chances of winning, you need not find the individual probabilities in this column.

Key Question

Do you have any shortcuts for finding the probability of the Good Guys winning?

Graphing the Games

Intent

This activity lays the groundwork for the definition of the binomial distribution.

Mathematics

Graphing the Games asks students to make two bar graphs, one showing the probabilities for each of the eight possible outcomes for the Good Guys' seven remaining games and one showing the probabilities for the eight possible outcomes of seven flips of a coin. They are then asked to compare the two graphs.

After the class identifies the common features of the two situations in the activity, the binomial distribution is defined. The discussion emphasizes that the binomial distribution, like the normal distribution, describes a family of situations. The shape of the distribution is determined by both the number of repetitions of the event and the probabilities of the two outcomes.

Progression

This activity provides the foundation for the binomial distribution, which is then applied in the next activity, *Binomial Probabilities*.

Approximate Time

5 minutes for introduction

25 minutes for activity (at home or in class)

25 minutes for discussion

Classroom Organization

Individuals, followed by whole-class discussion

Doing the Activity

We recommend that you select two students and have one prepare a transparency of the graph for Question 1 and the other prepare a transparency of the graph for Question 2. It will be helpful if they agree on scales for the two graphs so that the visual comparison will be clearer. (As an alternative, you can have them each prepare graphs for both problems.)

Discussing and Debriefing the Activity

Begin the discussion by having a display of the graphs from Questions 1 and 2. Have presenters explain how they found the probabilities. Emphasize the role of the binomial coefficients in computing the probabilities.

The graphs should look something like these:

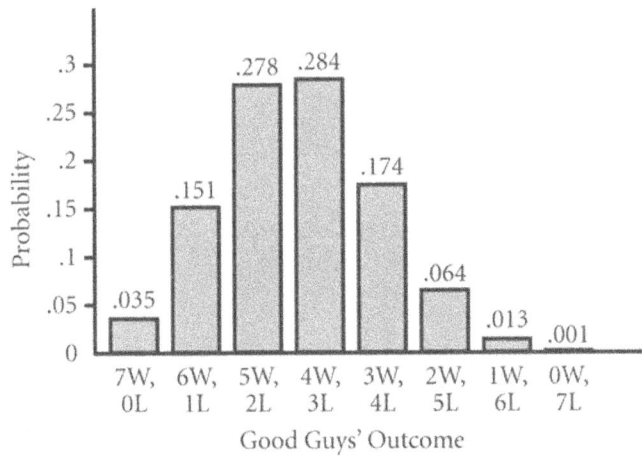

Good Guys' Outcome

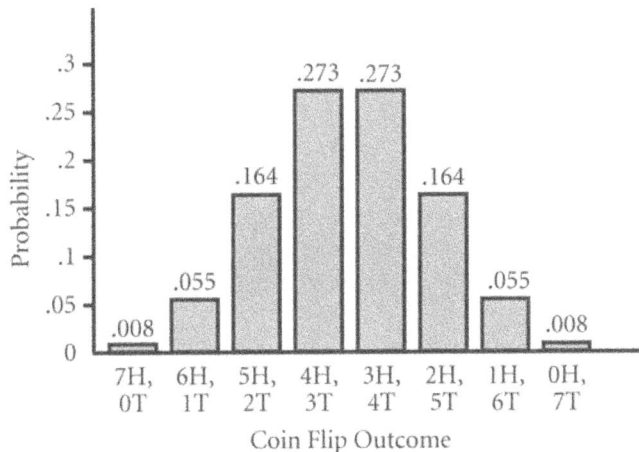

Coin Flip Outcome

Note: In the first graph, the last bar may be almost invisible, because the Good Guys' probability of losing all seven games is so small.

Question 3

Let students share ideas about how the graphs are the same and how they differ. Be sure to bring out these two key features.

- Both graphs are highest "in the middle" and lowest "at the ends."
- The coin flip graph is symmetric, while the Good Guys' graph is "shifted left."

Students might also comment that the coin flip graph resembles the normal distribution. If this comes up, tell the class that if the number of coin flips were greater, the resemblance would be even stronger.

The Binomial Distribution

Tell students that situations like the two in this activity come up often. Ask, **What key features do the two situations have in common?** Use their comments to focus on these two elements:

- There is some "event" with two possible outcomes.
- The "event" is repeated some number of times, but the probabilities of the two outcomes do not change. (That is, each occurrence of the event is *independent* of the previous occurrences.)

Ask, **What other situations fit this model?** For example, students might mention the scenario in *Don't Stand for It* or situations from outside the unit.

Tell students that the set of probabilities that goes with situations like these is called a **binomial distribution.** Clarify as needed that a distribution is not a single probability, but a description of the set of probabilities that go with all possible outcomes for a situation.

Ask, **What other 'distribution' have you studied?** If needed, remind them of the *normal* distribution, and point out that the normal distribution is also represented by a graph. (You might note that the normal distribution is a continuous graph while the binomial distribution is a discrete graph.)

Bring out that the binomial distribution and the normal distribution are each a *family* of distributions, dependent on certain parameters:

- For the binomial distribution, the specific set of probabilities depends on the number of times the event is repeated and on the probabilities for each event.
- For the normal distribution, the specific set of probabilities depends on the mean and the standard deviation.

For instance, students should see that the possible outcomes for the Bad Guys also follow a binomial distribution but that the distribution for the Bad Guys is slightly different from that for the Good Guys.

Ask, **From what does the binomial distribution get its name?** Bring out that the associated probabilities involve the combinatorial coefficients, also known as the *binomial* coefficients.

Binomial Probabilities asks students to find a general formula for the probabilities associated with the binomial distribution.

Key Questions

What key features do the two situations have in common?

What other situations fit this model?

What other 'distribution' have you studied?

From what does the binomial distribution get its name?

Binomial Probabilities

Intent

In this activity, students develop a formula for the probabilities associated with the binomial distribution.

Mathematics

Binomial Probabilities asks students to apply the binomial distribution to a situation with a large number of trials, then construct a formula for the probability of getting r successes in n trials in a situation having a binomial distribution, where the probability of success in each trial is p.

Progression

Students work through one specific example with a large number of games from the situation in the central unit problem, then construct a formula.

Approximate Time

20 minutes for activity (at home or in class)

10 minutes for discussion

Classroom Organization

Individuals, followed by whole-class discussion

Doing the Activity

Students should be able work on this activity independently with no introduction.

Discussing and Debriefing the Activity

Have a volunteer present Question 1. The presenter will probably give the answer as $_{162}C_{100} \cdot .62^{100} \cdot .38^{62}$. Elicit an explanation for each element of this expression. (For instance, the presenter should explain that if the Good Guys win 100 games, then they lose 62 games.)

Have students find the numerical value of this probability using a graphing calculator. (The value is approximately .064.)

Question 2

Have another student present Question 2. Be sure to go over the two issues raised in the hints:

- If the probability of success for each event is p, then the probability of failure for each event is $1 - p$.

- If the number of successes is r, then the number of failures is $n - r$.

With these two points clarified, students should see that the probability of exactly r successes is $_nC_r \cdot p^r \cdot (1 - p)^{n-r}$.

The formula itself is perhaps less important than the recognition that an entire family of problems can be analyzed using essentially the same reasoning.

Pennant Fever Portfolio

Intent

In this activity, students complete their unit portfolio.

Mathematics

For the portfolio, students write a cover letter that summarizes what they have learned in the unit and how it related to the unit problem. They also select activities from their completed work that were particularly important in developing the key ideas of the unit.

Discussion of the completed cover letters can serve as a starting point for a class review of the unit.

Progression

Preparing the portfolio gives students an opportunity to review the topics that were introduced in this unit and to reflect upon how they relate to one another.

Approximate Time

5 minutes for introduction

45 minutes for activity (at home or in class)

15 minutes for unit reflection

Classroom Organization

Individuals, followed by whole-class discussion

Doing the Activity

Students work independently writing their cover letter and selecting portfolio activities that reflect the mathematical ideas learned in the unit.

Discussing and Debriefing the Activity

Have volunteers share their portfolio cover letters as a way to start a discussion to reflect on the unit.

Blackline Masters

Playing with Probabilities

Choosing for Chores

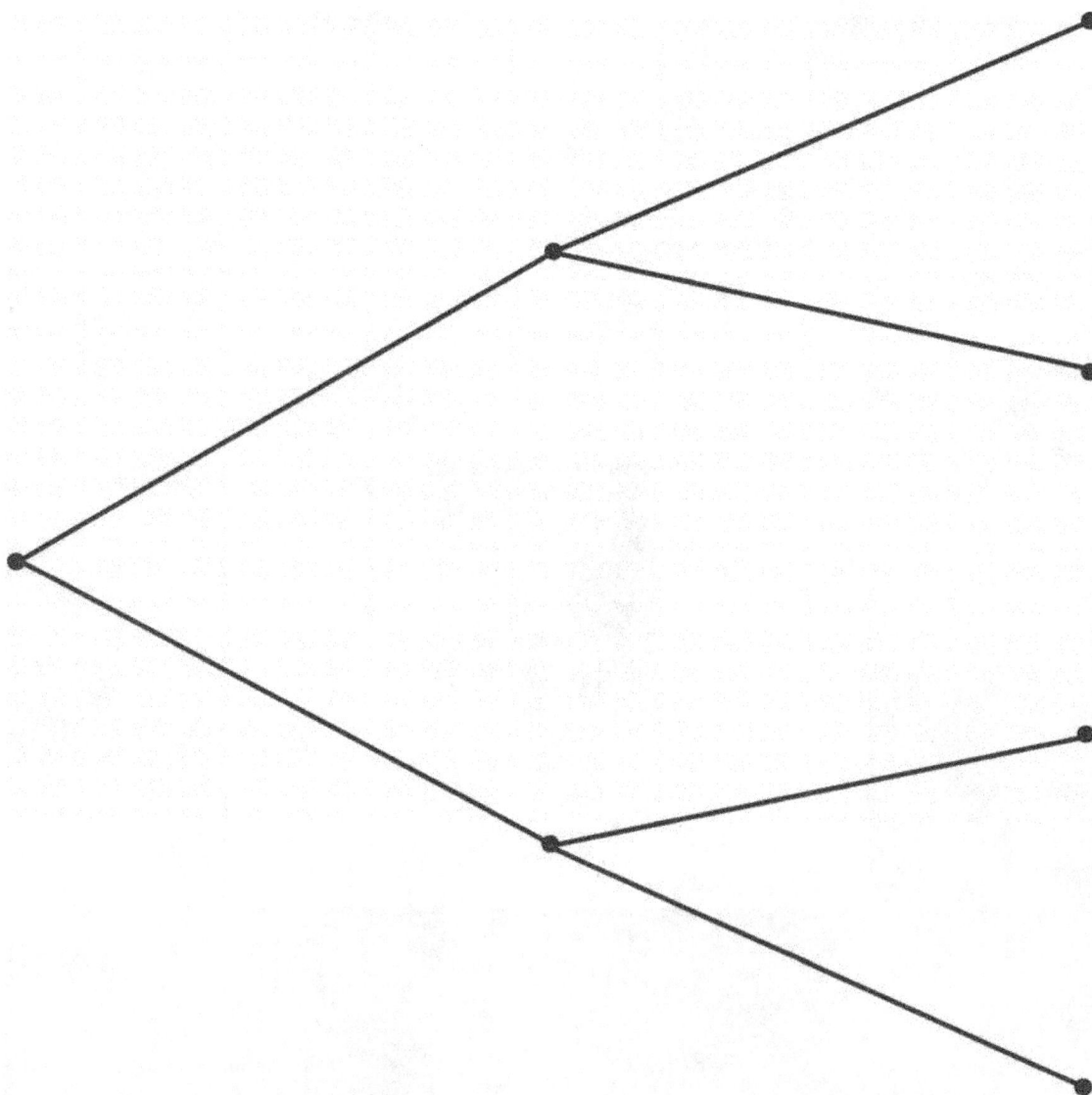

Possible Outcomes

Good Guys' record for the final seven games

	7−0	6−1	5−2	4−3	3−4	2−5	1−6	0−7
7−0								
6−1								
5−2								
4−3								
3−4								
2−5								
1−6								
0−7								

Bad Guys' record for the final seven games

¼-Inch Graph Paper

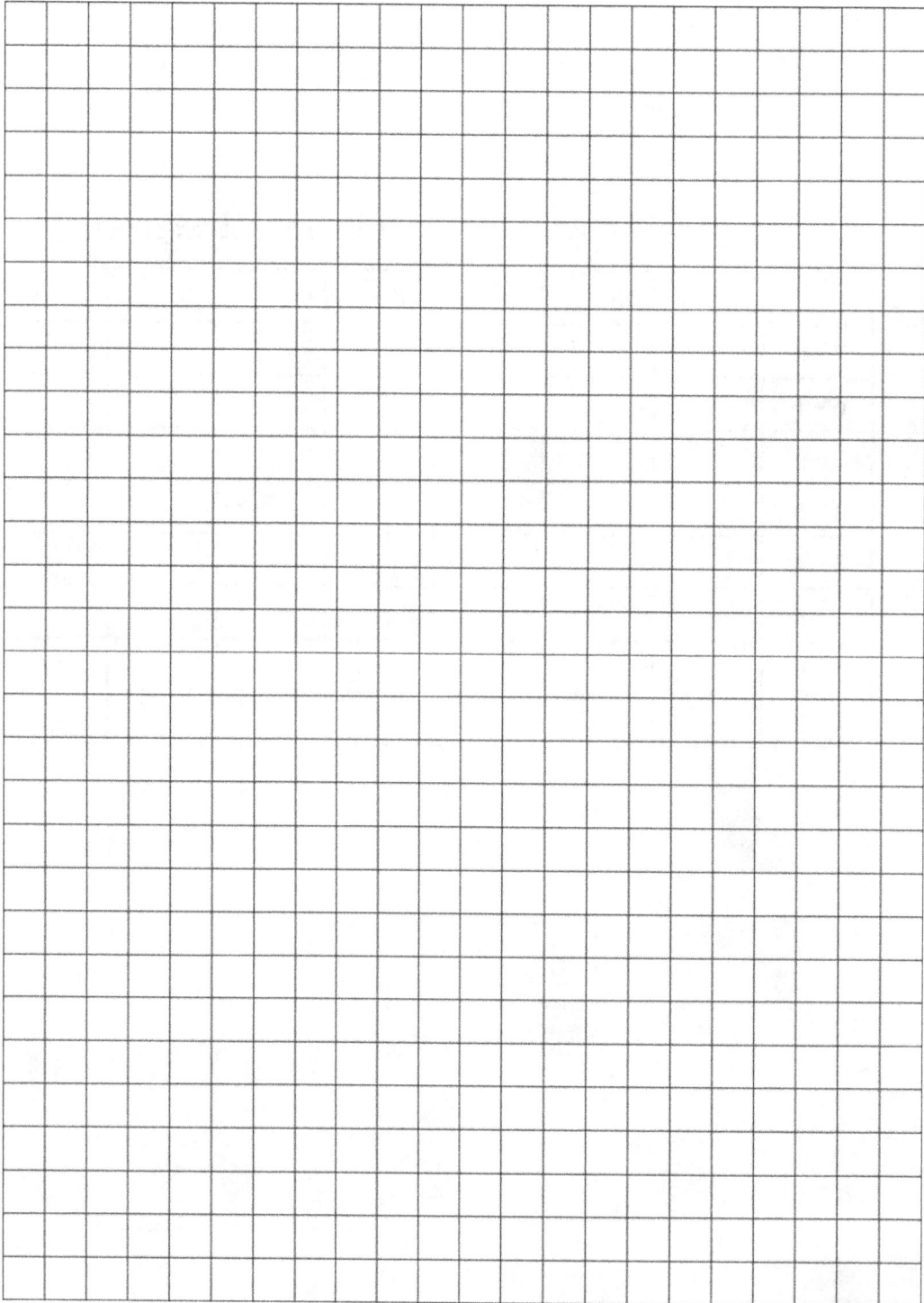

1-Centimeter Graph Paper

1-Inch Graph Paper

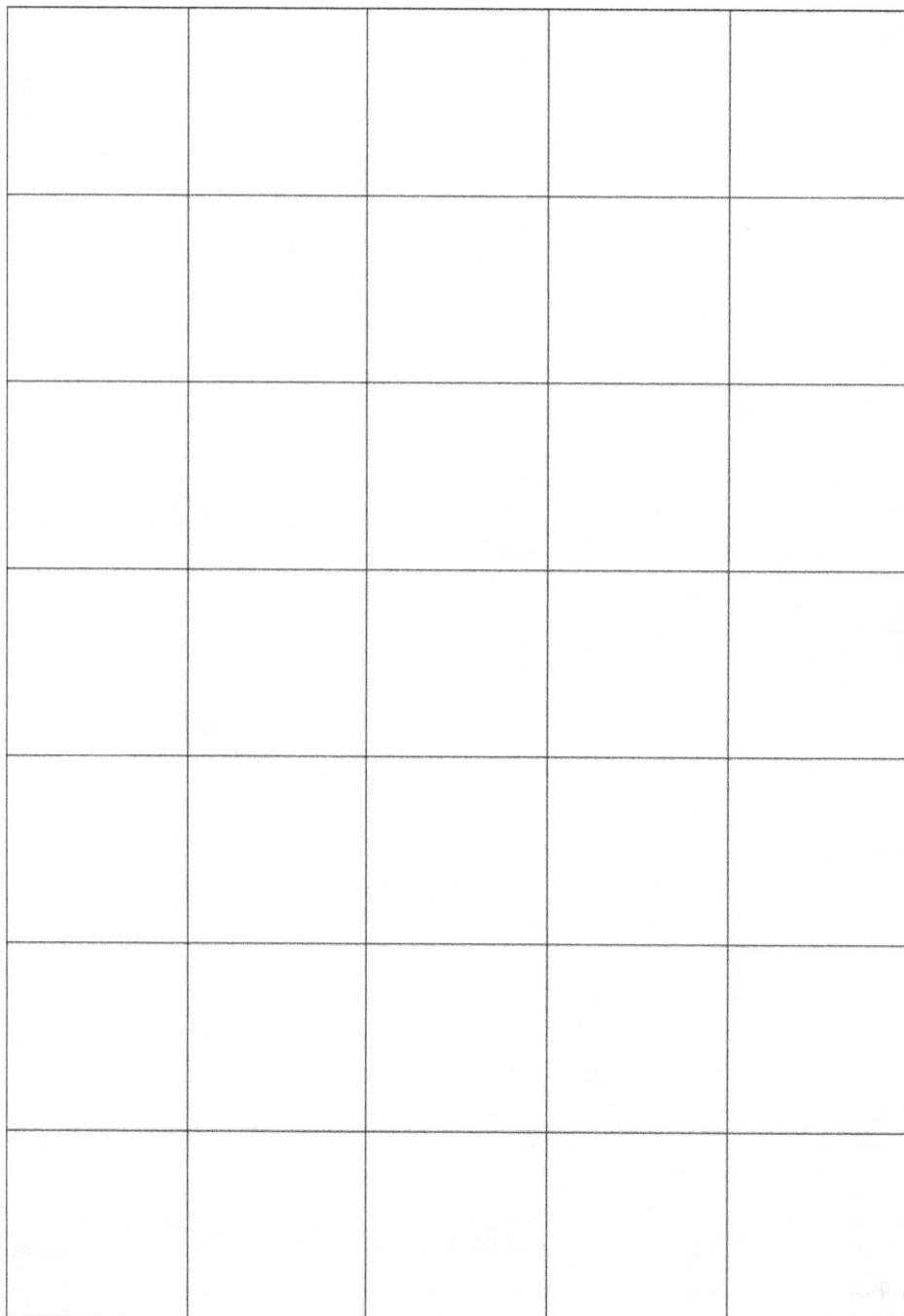

Assessments

In-Class Assessment

Janice and Lori have been hard at work all year, but each of them has also been daydreaming and planning for her upcoming vacation. Each will get five days off this year.

1. Janice's idea of a great vacation is to read a different mystery novel each day. She knows of 16 different ones she would like to read, but she'll only have time for 5. How many choices does she have about which 5 books to check out of the library? Explain your reasoning.

2. Lori has a different idea of a great vacation. She is an avid biker and has bought a book with ten day-long bike trips in her area. She'll be able to do only five of these. She needs to decide which trip to do on Monday, which to do on Tuesday, and so on. (Se won't repeat the same trip twice.) How many different schedules does she have to choose from? Explain your reasoning.

3. Lori is also really into all of the special clothes that go with biking. She has 8 different pairs of biking shorts and 12 different biking shirts. Because she gets very dirty biking, she'll have to wear different clothes each day. (She won't do any laundry on vacation, so no item will be worn more than once.)

   Lori doesn't want to waste time in the morning making decisions about what to wear, so she decides to label her outfits in advance: the Monday shorts, the Monday shirt, the Tuesday shorts, the Tuesday shirt, and so on. How many choices does she have for different ways to plan her wardrobe? Explain your reasoning.

Take-Home Assessment

Part I: Same Suit

If you are dealt five cards from a regular 52-card deck, what is the probability of getting all five cards of the same suit? Explain your answer.

Part II: Rained Out

The Good Guys have a series of five games to play, with one game each day from Monday through Friday. If it rains on the day of a scheduled game, that game is canceled. There is a 30% probability of rain each day.

How likely is it that the Good Guys will get to play at least two of the five games? Explain your answer.

Part III: $_nC_r$

Write a newspaper article about $_nC_r$. Most people have never heard of it. Describe what it is, how it is calculated, how it is useful, and anything else you think people might be interested in.

IMP Year 3
First Semester Assessment

I. Once Upon a Time . . .

Imagine that you are Madie or Clyde. You've grown old and are telling your grandchildren the story of the orchard hideout. You've described the arrangement of trees in the original orchard (with a radius of 50 units) and told them the basic facts that you knew at the start.

- The circumference of the newly planted trees
- The fixed amount by which the cross-sectional area of the trees grew each year
- The distance between the centers of adjacent trees

Of course, the grandchildren have heard the story before, and they remember that it took about 11 years, 9 months for the center to become a true orchard hideout. What you want to do is impress them with how well you and your partner analyzed the problem back then.

Write a description of how the analysis worked. Don't get bogged down in the specific numbers, because you don't have pencil and paper handy, and the youngsters are more interested in the big ideas anyway!

II. Road Building

The highway department is planning a road that will go through the town of Coldwater. The town of Hot Springs is 13 miles due south of Coldwater, and the town of Warm Rock is 18 miles due east of Hot Springs.

1. Sketch a diagram showing the relationship between the three towns. (Treat each town as a single point.)

The mayors of Hot Springs and Warm Rock both want this new road through Coldwater to go straight through their towns as well. Unfortunately, the highway department can afford to build only one road.

The road must go through Coldwater and must be straight, so a compromise route is needed. The mayors of Hot Springs and Warm Rock agree to support the project if this condition is met:
The distance from Hot Springs to the new road must be the same as the distance from Warm Rock to this road.

They also insist that the road should not be parallel to the route from Hot Springs to Warm Rock.

2. a. Add a dotted line to your sketch from Question 1 to show where the new road must go, and explain your reasoning.

    b. Find the distance from Hot Springs to the road, to the nearest tenth of a mile, and explain your work.

III. Equation Time

Solve this system of equations, and explain your work.

$$7r + 6s = 6$$
$$5r - 4s = 25$$

IV. The Third Dimension

This system of linear constraints in three variables defines a feasible region.

I	$2x + y + z \leq 20$
II	$3x + 4z \leq 13$
III	$y + 2z \leq 9$
IV	$x \leq 9$
V	$y \leq 6$
VI	$x \geq 0$
VII	$y \geq 0$
VIII	$z \geq 0$

Give *a general outline* of how to find the point in this feasible region where the function $2x + y + z$ has its maximum, and explain the geometric reasoning behind your method.

V. Solving with Matrices

Consider this system of linear equations.

$$3a + 2b - c + d = 1$$
$$2a - b + 4c + 2d = -2$$
$$-4a + 3c - 3d = -6$$
$$a + b + c + d = 3$$

1. Write a matrix equation that is equivalent to this system.

2. Solve the system using matrices on a graphing calculator, and show your solution.

3. Discuss the relationship between the matrices and the equations and the properties of matrices that allow you to use them to solve systems of linear equations.

I.   *Spilt Milk*

You've automated your dairy farm so that all the cows are milked by milking machines, and the milk all flows into one giant cone-shaped container. At the start of milking time, the container is empty, and as the milk flows in, the level in the container rises. Milking starts at 5:00 a.m. and continues through the day. (The cows are not all milked at the same time.)

After studying your cows and using some geometry, you've figured out that at t minutes after 5:00 a.m., the milk in the container will have risen to a level of $\sqrt[3]{2000t}$ centimeters.

1. During the hour from 7:00 a.m. until 8:00 a.m., what is the *average* rise per minute in the height of the milk? (Give your answer to the nearest 0.001 cm/min.)

2. At what rate is the milk level rising at 8:00 a.m.? (Again, give your answer to the nearest 0.001 cm/min.)

3. At what time of day will the milk level reach 100 centimeters?

II.   *Darts*

Consider a square dartboard with a circle inscribed in the square, as shown here.

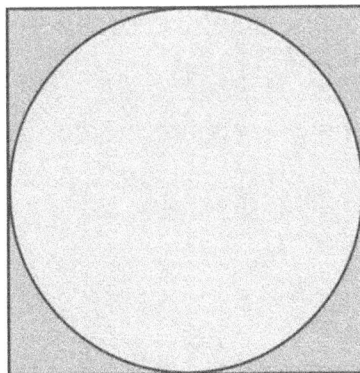

Suppose that according to the rules, if your dart lands inside the circle, you win, and if the dart lands outside the circle, you lose. Assume that you always hit the dartboard and that each point of the square is equally likely to be hit.

1. If you throw one dart, what is your probability of winning? Explain your answer, giving the probability to the nearest hundredth.

2. Suppose you throw seven darts. What is the probability that you will win at least four times? Explain what method you use to find the answer and why the method works. Again, give the probability to the nearest hundredth.

III. Ferris Wheel Fence, Revisited

It's time to look back at the problem of the fence around the amusement park, from *High Dive*.

As you may recall, Al and Betty are riding on a Ferris wheel. This Ferris wheel has a radius of 30 feet, and its center is 35 feet above ground level. There is a 25-foot-high fence around the amusement park, but once you get above the fence, there is a wonderful view.

What percentage of the time are Al and Betty above the level of the fence?

IV. Opposite Angles

You have learned these formulas involving trigonometric functions:

$$\cos(-\theta) = \cos \theta$$
$$\sin(-\theta) = -\sin \theta$$

Explain each of these formulas in several ways:

- In terms of the Ferris wheel
- In terms of the graphs of the sine and cosine functions
- Using numerical examples

You can use these graphs of sine and cosine in your explanation:

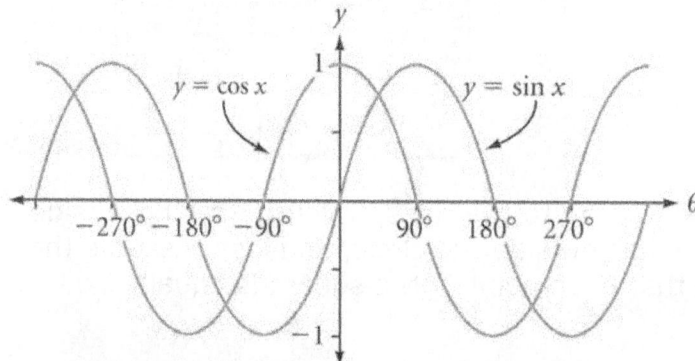

You can also use this diagram to represent a Ferris wheel:

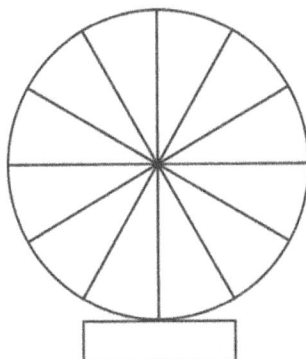

Pennant Fever Calculator Guide for the TI-83/84 Family of Calculators

During the course of this unit, students will extend their understanding of probability concepts and of systematic counting methods. *Pennant Fever* emphasizes techniques for determining theoretical probability—one way in which you will find the graphing calculator useful in these situations because it has a random number generator that can be used to simulate random events. Using this feature combined with the calculator's programming capability allows you to create and run simulations. Thus, you can run a large number of experiments to observe the likelihood of an event. These ideas were first explored in the Year 1 unit *The Game of Pig.* Also, as students explore ideas for modeling probability, the calculator's graphing capability will allow them to formulate and test algebraic conjectures along with their graphic representations.

The calculator will also be used in this unit to determine values for permutations and combinations.

The following discussion provides several opportunities to challenge students to explore and extend their mathematics and graphing calculator techniques.

Choosing for Chores: Some students may begin this activity by creating an experiment to simulate the situation. For example, they could place two purple tiles and three green tiles to represent the spoons in a paper bag. They could then select two tiles and see if they match. For this problem, the first tile drawn should not be put back in the bag.

The graphing calculator has a random number generator that can also simulate random events. The calculator must be set up to correctly simulate a situation, just as the paper bag was set up.

The Calculator Note "Random Numbers on the Calculator" gives basic tips on using this random number generator. The Calculator Note "Simulating Choosing for Chores" provides a program that simulates the spoon situation. You might wish to remind students about the simulations they programmed in the Year 1 unit *The Game of Pig.* You can use "Simulating Choosing for Chores" with students who show an interest or as an extension to the classwork. Or, challenge students to write their own programs to simulate the situation.

POW 12: Let's Make a Deal: As students design a simulation for the scenario in this POW, they may recall simulating random events using the random number generator on their calculator. Interested students can review the Calculator Note "Random Numbers on the Calculator."

The host must develop a random method for selecting the winning door. The command **iPart(3*rand)+1** will randomly select integers from 1 to 3. Press ENTER several times to get numbers like these:

```
iPart(3*rand)+1
                1
                1
                2
                3
                2
```

The supplemental problem *Programming a Deal* can be assigned to students interested in programming the calculator. Further comments and suggestions can be found in *Appendix A* in this section of the guide. You might have students start with *Simulating Choosing for Chores* in this guide, or you could challenge them to simulate that situation without the instructions.

The Real Birthday Problem: Multiplying the fractions involved in the birthday problems on the TI calculator can become cumbersome. For example, this screen shows $\left(\frac{364}{365}\right)\left(\frac{363}{365}\right)\left(\frac{362}{365}\right)\cdots\left(\frac{355}{365}\right)$. It's not easy to read!

```
(364/365)(363/36
5)(362/365)(361/
365)(360/365)(35
9/365)(358/365)(
357/365)(356/365
)(355/365)
        .8588586217
```

```
(364/365)(363/36
5)
        .9917958341
Ans*362
        359.0300919
Ans/365
        .9836440875
```

It is helpful to calculate each "additional person" from the previous results. For example, to get the probability of no match for four people, multiply the probability for three people by 362 and then divide that result by 365.

The Good and the Bad: As students continue to work with probabilities, they may ask to how many decimal places to round answers. You can help students avoid this issue by having them keep track of calculations using the last-answer (**ANS**) function of their calculator. In the screen here, the first line shows the probability that the Good Guys will win six of their seven games. In the second line, that result is multiplied by the probability that the Bad Guys will win six of their seven games.

```
7*.62^6*.38
        .1510886267
Ans*7*.60^6*.40
        .0197377347
```

More Cones for Johanna: To calculate the factorial of a number, first enter the number on the home screen, for example, 6. Then press MATH, arrow over to **PRB**, and select **4:!**.

```
6!
                   720
■
```

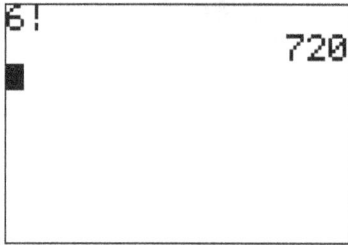

POW 13: Fair Spoons: The notations $_nP_r$ and $_nC_r$ are introduced along with this POW. It is most important that students relate these notations to the types of counting problems they have been doing; you should decide when it is appropriate to move beyond emphasizing the notation and the distinctions between the methods of counting. In the meantime, students may find and learn to use the calculator commands on their own.

POW 13: Fair Spoons could be considered an extension of the activity *Choosing for Chores.* With that in mind, students may wish to expand the program in the Calculator Note "Simulating Choosing for Chores" for their work on this POW.

Students may be interested in further extending *POW 13: Fair Spoons* by investigating patterns among the possible combinations of spoons for which the probability of a match is exactly .5. They may have noticed patterns from the results they generated or the list created in class.

If a general formula is developed (an example is in the *Teacher's Guide*), it can be set equal to .5 to create an equation that describes all spoon combinations that fit the criterion. Encourage students to use this equation to write a calculator program or to use the calculator's function table to generate a list of spoon combinations.

What's for Dinner?: Encourage students to figure out how to directly find $_nP_r$ and $_nC_r$ values using their calculator; they can likely find out how to do this on their own or with the help of a calculator guidebook. The Calculator Note "Combinatorics on the Calculator" gives detailed instructions that you may choose to give to students as a reference later on.

In the discussion following *What's for Dinner?,* students might be ready to develop the generalization $_nC_r = {_nC_{n-r}}$. However, this topic can be delayed until *Combinations, Pascal's Way.* If you think it appropriate to discuss this generalization today, ask students to make conjectures about what combinatorial coefficient is numerically equal to $_{62}C_{17}$. Students should test their conjecture using the calculator's **nCr** command. The size of this number may spark discussion:

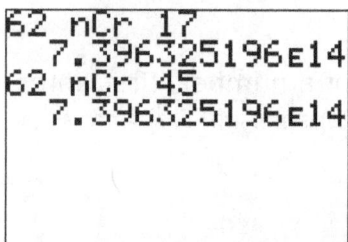

```
62 nCr 17
      7.396325196E14
62 nCr 45
      7.396325196E14
```

Binomial Powers: As students compare results of this activity, ask them how they can test to see if their expanded form of an expression is equivalent to the original expression. They may suggest looking at graphs or tables of the two expressions.

On the calculator (when dealing with only one variable), enter both functions in the Y= editor and then press either GRAPH or 2ND [TABLE].

```
Plot1 Plot2 Plot3
\Y1 ▤(X+1)²
\Y2 ▤X²+2X+1
\Y3=
\Y4=
\Y5=
\Y6=
\Y7=
```

```
 X     Y1     Y2
-2      1      1
-1      0      0
 0      1      1
 1      4      4
 2      9      9
 3     16     16
 4     25     25
Y1 ▤(X+1)²
```

Combinations, Pascal's Way: Rows or "columns" of Pascal's triangle can be found using the calculator's table function along with the **nCr** command. This might be an interesting thing to bring up at the end of today's class by asking, "What if you used **nCr** in the Y= editor?"

In **Y1**, enter **5 nCr X**. Press 2ND [TABLE].

```
 X     Y1
 0      1
 1      5
 2     10
 3     10
 4      5
 5      1
 6      0
Y1 ▤5 nCr X
```

What do you think will result if you enter **Y1 = X nCr 2**, and press 2ND [TABLE]?

Graphing the Games: This activity asks students to draw bar graphs that show the probabilities of outcomes that are binomial distributions. The calculator has the ability to create bar graphs. The Calculator Note "Graphing Probability Distributions" provides instructions for how to use this calculator feature. You may wish to provide this to students as an extension.

Binomial Probabilities: Once students have a sense of the general concept of a binomial distribution, you might want to show them how to find the binomial distribution probabilities on the calculator. To find the probabilities for the Good Guys, begin in the home screen, then press 2ND [DISTR]. Highlight **0:binompdf(** and press ENTER. Press 7 (because they play seven games), press , , and then press . 6 2 (because that is their probability of success), and press). If you now press ENTER, you get a list of the probabilities for each possible outcome. Press STO→ 2ND [L2] ENTER to store the eight resulting probabilities to List 2.

Assessments: Students should have access to graphing calculators during the in-class assessment.

Supplemental Activity—Programming a Deal: This activity challenges students to write a program to simulate the situation. If the results of many simulations on the calculators are near this situation's theoretical probabilities, students' programs likely are set up well. Students should be able to describe why their programs accurately model the situation and use the correct probabilities. If programs are not functioning, or are yielding surprising results, you might have students print their code using a computer with TI Connect (available free at education.ti.com). This will help in debugging.

Random Numbers on the Calculator

Many probability problems involve the idea of picking something at random. In the simplest case, each possible result is equally likely.

Your calculator has a feature called a *random number generator,* which you can use to simulate probability experiments. You may recall working with the random number generator feature in the Year 1 unit *The Game of Pig.*

The **rand** command generates a random number between 0 and 1. Begin in the home screen, press MATH, highlight **PRB**, select **1:rand**, and press ENTER. The **rand** command appears in your home screen. Press ENTER again, and the calculator will display a decimal between 0 and 1 (it is very unlikely to be the number shown here). Press ENTER repeatedly to generate more random numbers.

```
MATH NUM CPX PRB        rand
1 rand                          .1881244025
2:nPr
3:nCr
4:!
5:randInt(
6:randNorm(
7:randBin(
```

Simulating Rolling a Die

Random numbers between 0 and 1 can be used to simulate probability experiments. To model rolling a six-sided die, let random numbers between 0 and 1/6 (from 0 to 0.1666666666) represent rolling a 1, random numbers between 1/6 and 2/6 (from 0.1666666667 to 0.3333333333) represent rolling a 2, and so on.

Using Alternate Techniques to Simulate Rolling a Die

Reading all the decimals from the calculator can be tricky. Instead, enter **iPart(6*rand)+1** to get integers from 1 to 6, and press ENTER. (To find **iPart**, press MATH, highlight **NUM**, and select **3:iPart**.) Investigate what each part of this command does.

Another option is to use the **randInt** command. Press ENTER, highlight **PRB**, and select **5:randInt(**. Complete the command with the range from which you wish to select integers. Adding a third number defines how many random integers to generate.

```
iPart(6*rand)+1         randInt(1,6)
                5                          4
                2       randInt(1,6,5)
                4              {4 1 4 5 3}
                2
                4
```

Simulating Choosing for Chores

The activity *Choosing for Chores* gives a scenario in which Scott picks two spoons from a bag containing two purple and three green spoons. If the spoons he picks are the same color, he washes dishes. If they are a different color, he dries the dishes.

The instructions here describe a calculator program, called CHORES, that simulates this situation. A calculator running this program can quickly produce many random outcomes of the game. The display here shows the outcomes of three simulations.

```
PrgmCHORES
1 EACH
              Done
2 GREEN
              Done
2 GREEN
              Done
```

To enter the program into your calculator, start a new program, name it CHORES, and enter the instructions from the column on the left. The column on the right explains the function of each programming instruction.

Instruction	Explanation
:0→G	Stores **0** in a cell labeled **G**. This cell will be increased by one for every green spoon drawn. Use STO to enter the → symbol.
:(iPart(5*rand)+1)→S	Chooses a random integer between 1 and 5 and stores this number in a cell labeled **S**. To find **iPart**, press MATH, highlight **NUM**, and select **iPart**.
:If S>2	If the condition in an "If" instruction is true, the calculator carries out the next instruction. If the condition is false, the calculator skips the next instruction. Find **If** in the PRGM **CTL** menu. Find **>** in the 2ND [TEST] menu.
:Goto 2	If the random number **S** was 3, 4, or 5 (a 60% chance), Scott has drawn a green spoon first and will draw again—the program proceeds to label 2 later in the program. Find **Goto** in the PRGM **CTL** menu.
:(iPart(4*rand)+1)→S	The program gets to this line if a purple spoon was drawn first (because the condition S > 2 failed). Now, four spoons—one purple, three green—remain in the bag. Thus, the calculator chooses a random integer between 1 and 4 and stores this number in cell **S**, with the number 1 representing the purple spoon.
:If S>1	

:G+1→G	If the random number **S** was 2, 3, or 4, Scott drew a green spoon second. The cell **G** counts how many green spoons were drawn.
:Goto 9	The simulation proceeds to label 9 later in the program, where the calculator reports the results of the two spoons drawn.
:Lbl 2	Places a label that is proceeded to when a green spoon is selected first. Press PRGM and scroll down to select **Lbl**.
:G+1→G	Increases **G** by one to indicate a green spoon was drawn first.
:(iPart(4*rand)+1)→S	Now, four spoons remain in the bag—two of each color. Thus, the calculator chooses a random integer between 1 and 4 and stores this number in cell **S**, and the values 1 and 2 will represent the purple spoons.
:If S>2	
:G+1→G	If the random number **S** was 3 or 4, Scott drew a green spoon second. The cell **G** counts how many green spoons were drawn.
:Goto 9	The simulation proceeds to label 9.
:Lbl 9	This label is proceeded to from various points in the program. Instructions after this point report the outcome of the simulation.
:If G=0	Find the = sign by pressing 2ND [TEST].
:Disp "2 PURPLE"	Displays **2 PURPLE** if **G** is 0. Find **Disp** in the PRGM I/O menu. Press ALPHA [_] (found above 0) to enter the space.
:If G=1	
:Disp "1 EACH"	Displays **1 EACH** if **G** is 1.
:If G=2	
:Disp "2 GREEN"	Displays **2 GREEN** if **G** is 2.

If you like, you can modify this program by adding to or changing the display lines. You could also allow the user to input the number of green or purple spoons to be chosen from. It may also be helpful to have the program run repeatedly and then display the cumulative results.

Calculator Notes for the TI-83/84 Family of Calculators

Combinatorics on the Calculator

Your calculator will directly calculate values of $_nP_r$ and $_nC_r$. Press MATH and highlight the **PRB** menu—you'll see the **nPr** and **nCr** commands.

```
MATH NUM CPX PRB
1:rand
2:nPr
3:nCr
4:!
5:randInt(
6:randNorm(
7:randBin(
```

However, if you immediately select either **nPr** or **nCr**, **Ans nPr** or **Ans nCr** will be displayed in your home screen.

```
Ans nCr



```

You must first enter the *n* value, where *n* represents the total number of objects. Then enter the **nPr** or **nCr** symbol. Complete the command with the *r* value (the number of objects being selected).

For instance, to compute the value of $_7C_2$ used in *Five for Seven,* first enter **7**. Next, press MATH and highlight **PRB**. Then highlight **3:nCr**, and press ENTER. To complete the command, enter **2** and press ENTER.

```
7 nCr 2
            21

```

Use a similar process to verify that 840 is the value of $_7P_4$ (used in Question 1 of *Who's on First?*).

Compare the values of $_5C_2$ and $_5P_2$. How are these values related? Is this relationship true for $_9C_2$ and $_9P_2$ also? What happens for larger values of *r*?

Graphing Probability Distributions

In *Graphing the Games,* you created a bar graph showing the probabilities of the different possible outcomes for the Good Guy's baseball team. To create the bar graph by hand, you calculated the probability for each outcome and sketched the bars. Your calculator can draw the bar graphs for you once you have the probabilities.

Begin by entering the number of possible wins, 0 through 7, in List 1 and the corresponding probabilities in List 2. To do this, press STAT, press ENTER to select **1:Edit...**, and then type values into the lists.

Next, press 2ND [STAT PLOT] ENTER to adjust the options for **Plot1** as shown here. Press 2ND [L1] or [L2] to enter **L1** or **L2**. Be sure to turn the plot on and highlight the bar graph option.

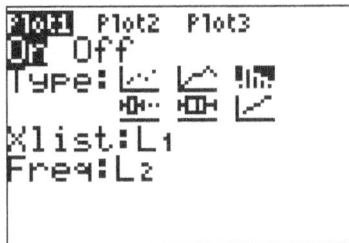

Next, press WINDOW to set the viewing window (or press ZOOM and select **ZoomStat**). Use the values shown here.

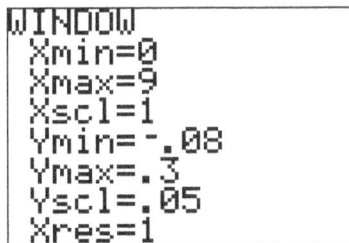

Finally, press GRAPH.

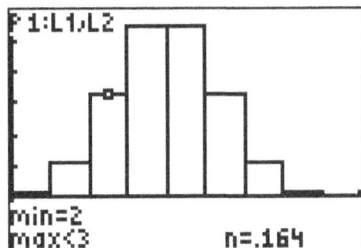

Try adjusting these steps to draw the coin flip graph. Also, look in the table of contents of your calculator guidebook to learn more about the instructions and commands used here.

www.ingramcontent.com/pod-product-compliance
Lightning Source LLC
Chambersburg PA
CBHW051344200326
41521CB00014B/2477